INTERNATIONAL CENTRE FOR MECHANICAL SCIENCES

COURSES AND LECTURES - No. 83

PETER W. BECKER

TECHNICAL UNIVERSITY OF DENMARK

AN INTRODUCTION TO THE DESIGN
OF PATTERN RECOGNITION DEVICES

COURSE HELD AT THE DEPARTMENT
FOR AUTOMATION AND INFORMATION
JULY 1971

UDINE 1971

SPRINGER-VERLAG WIEN GMBH

Originally published by Springer-Verlag Wien - New York in 1972

ISBN 978-3-211-81153-5   ISBN 978-3-7091-2650-9 (eBook)
DOI 10.1007/978-3-7091-2650-9

PREFACE

It is the purpose of this paper to introduce the reader to the fundamentals of pattern recognition theory. Having read the paper the reader will have gained an understanding of the problems inherent with design of pattern recognition devices sufficient to form an opinion on practical matters and to appreciate the current literature. The presentation of the theory is kept at an undergraduate level; for the sake of illustration two pattern recognition devices are discussed in some detail in Appendices 1 and 3. In the Addendum the answer is presented to the following, fundamental answer question in statistical pattern recognition: how does one find the multivariate probability densities which have a specified set of marginals?

Peter W. Becker

Udine, July 1971

GLOSSARY OF SYMBOLS

A   A class of patterns, Section 2.

a   Sample mean frequency for members of class A, Subsection 5.4.

B   A class of patterns, Section 2.

$B(j)$ The expected cost of classifying a C_j member with the minimaxing procedure, Subsection 6.3.

b   Sample mean frequency for members of class B, Subsection 5.4.

C_i Pattern class n°i of the N_c classes.

FOBW   Frequency of Occurrence of Binary Words, Appendix 4.

G_1 Average information about class membership given by Ξ_1; Subsection 5.4. Paragraph (4).

IP   Index of Performance, Section 6.

J_{AB} The symmetric divergence; Subsection 5.4, Paragraph (5).

K_{ij} The cost of classifying a member of class n°i as being a member of class n° j, Subsection 4.1, Paragraph (b).

$(K_i)_k$ The average cost incurred when all patterns represented by points in W_k are classified as C_i members, Subsection 6.1.

L_k The likelihood ratio for micro-region W_k, Subsection 6.4.

M_{AD} The number of Class A members in the design data group, Section 2.

M_{AT}	The number of Class A members in the test data group, Section 2.
M_{BD}	The number of Class B members in the design data group, Section 2.
M_{BT}	The number of Class B members in the test data group, Section 2.
MS	Measure of Separability, Subsection 5.4.
N_C	Number of pattern classes.
N_H	The largest number of attributes that can be implemented in the hardware (or software) realization of the pattern recognizer.
N_P	The number of attributes from which the set of p attributes is selected, Subsection 5.6.
P_K	The probability of a point falling in W_k , Subsection 6.1.
$P(i)_k$	The probability of a point representing a C_i member falling in W_k, Subsection 6.1.
$P_{pr}(i)$	The a priori probability of the appearance of a member of pattern class n° i, Subsection 4.1, Paragraph (a).
PRD	Pattern recognition device.
p	The number of attributes which is used, $\Xi_1,..,\Xi_p$, $p \leqslant N_H$, Figure 1.
q	The number of measurements , Υ_1 , ... , Υ_q , Figure 1.
R^*	Bayes' risk; Subsection 6.2.
$R(i)$	The expected cost of classifying a C_i member with Bayes' procedure, Subsection 6.2.
S	A Measure of separability, Subsection 5.4.

S$_H$ The number of hyperplanes, Subsection 7.4.

T   Thresholdsetting, Figures 2 and 3.

W$_k$ Micro-region n°k, Subsection 6.1.

Ξ Xi . The vector with the p coordinates $(\Xi_1,$ $\ldots, \Xi_i, \ldots, \Xi_p)$, Section 2.

Ξ_i Attribute n° i, Figure 1.

1. Introduction

1.1 The Purpose of the Paper

Within the last few years the interest in machines for recognition of patterns has increased sharply. This circumstance is for instance evidenced by the increasing number of conference sessions dedicated to the subject. Many papers in the area do, however, not receive the attentions from the engineering community they deserve; the reason is that the papers are concerned with details the importance of which can only be appreciated if the audience has some understanding of how a pattern recognition device is designed. (A "pattern recognition device" will in the following be abbreviated as PRD). It is the purpose of this paper to present a birds eye view of the field of PRD-design; it is hoped that the paper at the same time may serve as a primer for engineers who are becoming engaged in work with PRDs. Most of the published work on PRD-design has been concerned mainly with categorizer design methods (Sebestyen 1962; Nilsson 1965; Nagy 1968; Wee 1968; Ho and Agrawala 1968), in this paper the accent is on the design of the whole system.

The field of pattern recognition is new, and no all-comprehending theory is at present available. Time and again during the design of a PRD it becomes necessary for the designer to make educated guesses, as will be seen in the following. This fact may make the design of PRDs less of an exact science, but the importance of "the good hunch" adds a certain challenge to the work in this new field. For the purpose of this paper a class of patterns is defined as a set of objects which in some useful way can be treated alike. A number of assumptions have been made in the body of the paper. To make it clear when a restrictive assumption is introduced, the word assume will be underlined in each case. A number of important and readily available publications are listed in the bibliography. A comprehensive Pattern Recognition Bibliography has been compiled by Dr. C.J.W. Mason, it was published in the IEEE Systems, Science and Cybernetics Group Newsletter in 1970 and 1971.

1.2   The Two Phases in the Existence of a PRD

The existence of a PRD may be divided into two phases: the design phase (also called the learning phase), and the recognition phase. In the recognition phase the machine performs the following operation: when presented with an unlabelled pattern, the machine decides to which of some previously specified classes of patterns the given pattern belongs. One of the pattern classes could by definition contain "patterns not belonging to any of the other classes". The recognition phase is

is the "useful phase" where the PRD performs work by classifying patterns and thereby helps the user to make decisions under
conditions of uncertainity. The PRD acts as a (perfect or imper
fect "clairvoyant" who reduces the economic impact of uncertain
ty for the user. The value of the PRD's "clairvoyance" may be
computed (Howard 1967); it has a strong influence on the price
of the PRD and on the user's preference of one PRD over another.
Before the PRD can recognize patterns reliably, the PRD must
have been trained somehow to pay attention to the significant pat
tern attributes. This training is performed during the design
phase. The work described in this paper is concerned only with
the design phase.

1.3   Three Areas of Application

The application of PRDs fall essentially into
three areas.

1.3.1   Certain Acts of Identification

This is the area where the human specialists
are getting scarce or where they may be lacking in speed, accuracy or low cost. Some examples from this area will now be
mentioned; in each case PRDs have been reported to be success
ful or to hold much promise. Reading of alpha-numeric charac

ters which may be impact printed (Liu and Shelton 1966) or hand

printed (Brain et al. 1966; Genchi et al. 1968; Munson 1968 con

tains reviews of more than 40 papers): much effort has been ap-

plied in this area where commercial applications seem to be

abundant; a selection of papers on the reading of printed charac

ters has been indicated by the descriptor D_{10} in Minsky 1963,

a second selection appears in Kanal 1968, Part 1, the following

special issues should also be noted: IBM Journal of R and D,

vol. 12, n° 5, Sept. 1968, The Marconi Review vol. XXXII, n°

172, First Quarter 1969, and Pattern Recognition, vol. 2, n° 2,

n° 3, Sept. 1970. The automated reading of other kinds of charac

ters, e. g. , printed Chinese (Casey and Nagy 1966) or Japanese

(Sugiura and Tetsuo 1968) characters, has also been given atten

tion due to the increasing interest in the machine translation.

There is a growing need for medical mass-screening devices of

the PRD type; some steps in this direction have already been

taken with regard to the inspection of heart sound patterns (Young

and Huggins 1964) and electrocardiograms (Caceres and Dreifus

1970). Other examples of acts of identification are: detection of

cloud patterns from satellite altitudes (Darling and Joseph 1968),

surveying of earth resources from aerospace platforms (Fu et

al. 1969), automatic decoding of handsent Morse code (Gold 1959),

jet engine malfunction detection based on engine vibrations (Page

1967), detection of hyperthyroidism (Kulikowski 1970), speaker

identification (Kersta 1962; Dixon and Boudreau 1969; Das and

Mohn 1971) classification of white blood cells (Prewitt et al.,
1966; Ledley 1969; Ingram and Preston 1970), and the solving
of jig-saw puzzles (Freeman and Garder 1964). Two examples
of PRDs that perform acts of identification will briefly be described in Appendices 1 and 3.

1.3.2 Decisions in Regarding Complex Situations

There are a number of decisions making prob
lems having the following properties. (1) The number of possible
alternative decisions is relatively small, meaning that the num
ber of pattern classes N_c is relatively small. (2) The input on
which the decision must be based, meaning the unlabelled pattern that should be classified, is hard to evaluate; in other words,
the designer has real difficulties finding an effective set of pat
tern attributes. Three problems where the input has the form.
of a matrix will illustrate this type of problem: prediction of
tomorrow's weather from taday's weather map (Miller 1961;
Widrow et al. 1963), and the selection of a best move with a
given board position in checkers (Samuel 1963; Samuel 1967) or
or in chess (Newell et al. 1963; Newell and Simon 1965; Green
blatt et al. 1967; Good 1968). A number of problems concern-
ing business and military decisions are of the same "best move"
type.

1.3.3 Imitation of Human Pattern Recognition

There are a number of pattern recognition prob lems that are solved by humans with the greatest of ease and which so far have been largely impervious to solution by machine. Typical of such problems are the reading of handwritten material (Mermelstein and Eden 1964) and the identification of a per son from an ensemble of persons based on his handwriting (Eden 1962, Fukushima 1969) or appearance, (Tamura et al. 1971). Such a difficult pattern recognition problems are of interest not only for commercial reasons but also because their solution may shed light on cognitive processes. (Feigenbaum and Feldman 1963, Part 2; Lindgren 1965; Deutsch 1967; Dodwell 1970; Tsypkin 1968).

2. The Configuration of a PRD

Before the design of a PRD is discussed the structure of the finished product will be described briefly. Follow ing the usual system description (Marill and Green 1960), a pat tern recognizer consists of two parts, a receptor and a catego rizer, as shown in Figure 1; preliminary processing is disre- garded at this point, it will be discussed in Subsection 5.1. The receptor may also be referred to as the feature extractor or the characteristic calculator. The categorizer is also called the decision maker or the classifier. The functions of the two parts

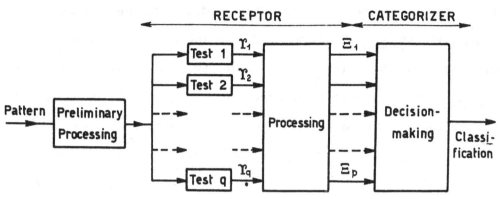

Fig. 1. Block Diagram of a Pattern Recognizer.

become apparent when it is considered how a PRD (in the recogni

tion phase) decides to which of several classes of patterns an un-

labelled pattern belongs. The number of pattern classes is called

N_C . In the receptor, the unlabelled pattern is exposed to a battery

of different tests. The test results constitute a set of numbers Υ_1,

$\Upsilon_2,...,\Upsilon_q$, these numbers are processed in some manner to yield the

set of numbers $\Xi_1,\Xi_2,...,\Xi_p$, that constitute the input to the categorizer.

The p numbers may be illustrated geometrically by a point, cal

led a pattern point, in a p-dimensional space, referred to as

pattern space or decision space; the pattern point is called Ξ =

= $(\Xi_1,...,\Xi_p)$. The p numbers are the values of the p attributes

in terms of which the unlabelled pattern is now described. The

concept "Attributes" has also been known in the literature under

the names: properties, features, measurements, characteris-

tics, observables, and descriptors. The word "attributes" is us

ed in this report because the other words have wider uses; con fusion seems less likely when the word "attribute" rather than one of the other words is selected to describe the attribute con cept. The categorizer stores information describing the rule ac cording to which decisions are made. In this report the rule is assumed to be non-randomized, meaning that the categorizer always makes the same decisions if the same pattern point $\Xi =$ $= (\Xi_1, ..., \Xi_p)$ is being categorized several times. The rule is equivalent to a decision function illustrated by a decision surface (also called separation surface) which partitions the p dimensional space. The decision surface may be implicitly defined as in the case of the machine "Pandemonium" (Selfridge 1959). With this machine first the N_C a posteriori probabilities of class membership are computed given the location of the pattern point after which the pattern is classified as belonging to the most probable class. The decision surface may also be explicitly de fined as a geometric surface, e. g., parts of three hyperplanes may constitute a decision surface as will be illustrated by an example in Figure 6. The decision surface which is stored in the categorizer partitions the pattern space into N_C or more compartments. During categorization it is decided to which of the N_C pattern classes the unlabelled pattern belongs by:

(1) finding the compartment in which the pattern point corresponding to the unlabelled pattern is located; the compartments are so located that points in the same compartment as nearly as possible represent members of the same class of patterns.

And

(2) announcing the name of the pattern class which pattern points prevail in the compartment; this could be done by lighting a particular indicator light. There is usually no indication of the level of confidence associated with the decision.

The design of a PRD is tantamount to a choice of tests, processing methods, and partitioning of the decision space. PRDs may be so designed that they always arrive at a classification based on the q measurements, they could, however, have a reject capability (Chow 1971). Most of the PRDs discussed in the literature perform all q tests on the unlabelled pattern. In cases where the tests are expensive the tests are performed one at a time and each new test is selected in the light of the results of the previously performed tests (much in the same manner the members of a panel show selected new questions); machines where such deferred-decision or sequential-detection techniques are used have been described in the literature (Selin 1965, Chapter 9; Chen 1966; Chien and Fu 1968; Nelson and Levy 1968; Mendel and Fu 1970, Chapter 2).

The designer's choice of tests, processing methods, and location of the decision surface is almost always influenced by constraints on time and funding, the weight and volume of the final implementation, etc..To appreciate how a designer may use the available techniques, it is instructive to summarize the factors that influence the design; this will be done in Section 4. As an

introduction, a block diagram will be used to illustrate the pro
cedure when a PRD is designed. A PRD does not necessarily
have to have the physical appearance of a special purpose com
puter; it could very well take the form of a program for a general purpose computer.

The problem of designing pattern recognizers may also be con
sidered as a problem in mathematical linguistics (Knoke and
Wiley 1967); this interesting approach is presently in the state
of development and will not be discussed further. The basic idea
is that any pattern can be reduced to a finite string of symbols
and that classification (or labelling) of such strings is similar
to mechanical translation of natural languages.

3. A Step by Step Description of the PRD Design Procedure

The Block Diagram illustrates the seven steps
in the design of a PRD. It is assumed in this section that a pat
tern always belongs to one of two classes, Class A or Class B;
the assumption of $N_c = 2$ is not very restrictive for reasons
which will be given later in Subsection 7. 4.

(1) First the designer will contact the specialists in the field, as
indicated by Block n°1. The designer wants to learn how in the
opinion of the specialists (i) the processes that generate the pat
terns may be modelled (Block n° 5) and how (ii) a library of representative patterns should be obtained (Block n°3).

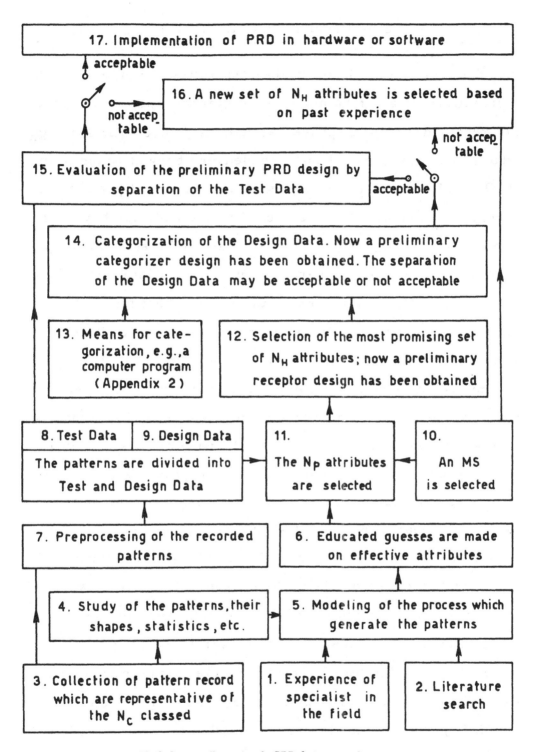

17. Implementation of PRD in hardware or software

↑ acceptable

not acceptable

16. A new set of N_H attributes is selected based on past experience

not acceptable

15. Evaluation of the preliminary PRD design by separation of the Test Data

acceptable

14. Categorization of the Design Data. Now a preliminary categorizer design has been obtained. The separation of the Design Data may be acceptable or not acceptable

13. Means for categorization, e.g., a computer program (Appendix 2)

12. Selection of the most promising set of N_H attributes; now a preliminary receptor design has been obtained

8. Test Data | 9. Design Data

The patterns are divided into Test and Design Data

11.

The N_P attributes are selected

10.

An MS is selected

7. Preprocessing of the recorded patterns

6. Educated guesses are made on effective attributes

4. Study of the patterns, their shapes, statistics, etc.

5. Modeling of the process which generate the patterns

3. Collection of pattern record which are representative of the N_C classed

1. Experience of specialist in the field

2. Literature search

Block diagram illustrating the PRD design procedure.

The data collection phase (Block n°3) is often expensive and demanding in time and funding. It is crucial that the greatest care be exercised during this phase because the future PRD is designed by generalizing from these recorded members of Class A and B. Quite often in practice, however, the library of patterns has already been obtained so the recording methods are beyond the control of the designer. Block n°4 illustrates a study of the pattern records by the designer. The study is not intended to be a painstaking effort, its purpose is to insure that more obvious pattern characteristics are detected and incorporated in the process model. The study begins with a visual inspection of the representative patterns. Next some simpler parameters, e.g. the waveform mean (Fine and Johnson 1965; Anderson 1965) may be studied for each waveform. Then the patterns are tested for hunches regarding their nature. E.g. the hypothesis that a waveform has been generated by a Gaussian process may be tested by the Chi-Squared method. It should be noticed that when a statistical method is selected for testing hypothesis, (Lehman 1959; Hald 1962) the method should be chosen in the light of what is known about the distribution of the variables and the number of samples available. If the level of measurement (Siegel 1956) is nominal, ordinal or interval (rather than ratio) the designer must use a non-parametric or rank test (Siegel 1956; Hajek and Sidak 1967). E. g. the hypothesis that the distribution of amplitude values is stationary can be tested with a Wald-Wolfowitz runs test

(Siegel 1956, Chapter 6). Block n° 2 illustrates a search through the pertinent literature by a librarian. Such a search is inexpensive and will often be found to give excellent returns. Based on the information from Block n° 1, 2 and 4 the designer decides on a model for the kind of process or processes that generated the Class A and Class B patterns, this step is illustrated by Block n° 5. The modelling problem has been discussed in the literature (Martin 1968; Klir and Valach 1968). E.g. if the members of a class are handprinted capital E letters the model may be: an essentially straight or vertical line from which at the approximate top, midpoint, and bottom, three, somewhat shorter, essentially straight, horizontal lines extend to the right; e.g. if the members of a class are time series generated by a stationary Markov process the model will include a transition matrix. The features that are believed to separate members of Class A from members of Class B should, of course, be related to the transition probabilities.

(2) Block n° 6 illustrates the most critical step in the PRD-design: the designer makes guesses on attributes, which he thinks will be effective in separating members of Class A from members of Class B. Let it be assumed that the designer selects N_p attributes. N_p is usually larger than the largest number of attributes the designer can use in practice, so the designer will later be faced with a selection problem. Although some general guidelines may be given for the selection of attributes (as will

be done in Subsection 5.2) the success of this step does largely hinge on the designers resourcefulness. The question of how the effectiveness of an attribute should be defined will be discussed in Subsection 5.3.

(3) The library of representative patterns is divided into two groups. The first group, the Design Data, is used to obtain a preliminary design for the PRD. The group contains M_{AD} members of Class A and M_{BD} members of Class B. Only this group will be considered in Steps n° 4, 5 and 6 in this section. The second group of patterns, the Test Data, is used in a later, computer simulated test of the designed machine. The Test Data group contains M_{AT} members of Class A and M_{BT} members of Class B. The division of the data is indicated by Blocks n° 8 and 9   How best to divide the available data has been discussed in Highleyman 1962. Before the data is divided it is usually exposed to some form of preliminary processing as indicated by Block n° 7; preliminary processing, will be briefly discussed in Subsection 5.1.

(4) The designer measures the value of each of the N_p attributes for each of the M_{AD} plus M_{BD} patterns (or at least a representative sample of the design patterns). The M_{AD} plus the M_{BD} sets of N_p attribute values each constitute the starting point for a search for a set of N_H (or fewer) attributes that will give adequate separation of the Class A and Class B members; N_H is the largest number of different pattern attributes the designer

can afford implement. Typically, N_H between 5 and 150. The search for an effective set of N_H attributes is tantamount to a design of the PRD by iteration as will become apparent during the remaining part of this section.

For each attribute the value of a suitable measure of separabil ity, MS, is computed. The value is an indication of the ability of the attribute, when used alone, to separate members of Class A from members of Class B. (A subjective estimate based on the histogram for Class A and Class B members would serve the same purpose but is impractical). Seven possible MS func tions are listed in Subsection 5.4. The attributes are ordered in an array according to their value of the MS. After the array has been established all the coming events in the PRD design effort have largely been determined. Next the designer selects N_H attributes from the N_p attributes in one of the ways described in Subsection 5.6. The selection of the N_H attributes con stitutes a first preliminary receptor design as indicated by Block n°12 in the block diagram.

(5) Any pattern from the library of representative patterns may be described by a point in N_H-dimensional space which as coordinates have the values of the N_H attributes. The M_{AD} plus M_{BD} points representing the members of Classes A and B constitute the input for the means for categorization by which the preliminary categorizer is designed. Block n°13 indicates the means for cate gorization; the topic of categorization will be discussed in Section

7. The mean for categorization is usually a computer program (plus matching computer) as described in Appendix 2; the mean could also be a trainable device (Block 1962; Brain et al. 1963; Widrow et al. 1963; Steinbuch and Piske 1963). The result of the categorizer design is in geometrical terms the location of a separation surface of given functional form e. g. a hyperplane. The surface separates the representative Class A points from the representative Class B points in an optimum manner according to some decision procedure (Mengert 1970). Four possible decision procedures will be described in Section 6. With each procedure is associated an index of performance, IP; e. g. if decisions are made so as to minimize the average cost of misclassifying a pattern this minimized cost, R^*, is a possible IP for the PRD. Usually some of the points will be misclassified. During the categorization procedure it should be noticed which attributes seem to contribute most to the separation. This important point will be discussed in Subsection 5.6. The computation of a preliminary categorizer constitutes Step n° 5. When the categorizer component values have been computed, a preliminary design for the pattern recognizer has been ontained. It consists of the preliminary receptor followed by the preliminary categorizer. Step n° 5 is illustrated by Block n° 14.

(6) Now the IP value of the preliminary PRD is determined and compared to the (customer-) specified value, IPS for the desired PRD. The preliminary design may be considered " not ac

ceptable" when $IP < IPS$. If so, new and better attributes are needed. Block n° 16 illustrates the generation of a new set of N_H
attributes by the designer based on his newly gained experience.
The only systematic method for generating new sets of promis
ing attributes is to the best of the authors knowledge the FOBW-
method (Page 1967, Becker 1968) described in Appendix 4. Each
time a new set of N_H more promising attributes has been select
ed, the new and presumably more effective receptor is tried out;
Step n° 5 is repeated. When the preliminary PRD design separates the design data in an "acceptable" manner, Step n° 6 has
been accomplished, this usually requires some iterations. The
number of attributes actually used in the design is called p as in
Fig. 1, $p \leqslant N_H$. In most practical cases $p = N_H$. $p < N_H$ occurs in the
rare cases where perfect categorization is achieved with fewer
than N_H attributes. Step n° 6 is carried out by the feedback
loop of Blocks n° 16, 10, 12 and 14.

(7) How well can the latest preliminary pattern recognizer, con
sisting of the preliminary receptor followed by the preliminary
categorizer, be expected to perform? This problem will be dis
cussed in Subsection 4.2. An estimate of the error probability
may be obtained by letting the preliminary pattern recognizer
separate the Class A and Class B patterns from the test data
group (Highleyman 1962 a). The estimate is obtained by taking
the following steps. First, the values of the p attributes that are
used in the latest of the preliminary receptor designs are com

puted for each of the M_{AT} plus M_{BT} patterns. Each of the test data patterns may now be described by a point in p-dimensional space. Secondly, the IP value for the test data is computed. The result may be taken as an estimate of the quality of the PRD design. Also, the confidence limits for the estimate should be computed, if possible, (Highleyman 1962 a). If the preliminary PRD is acceptable, the design phase is over; the design is considered final, and the PRD is realized in hardware as indicated in Block n° 17. If the preliminary design is not acceptable, the PRD design is modified by iteration so that it can separate the design data, with even greater accuracy. The modified PRD design is again used in Block n° 1(to separate the test data. If the separation by now is acceptable, the PRD design is realized in hardware as indicated by Block n° 17. If the separation of the test data is still found "not acceptable", the PRD is again modified by iteration so that it can separate the design data even better, etc. The underlying assumption is that the series of PRD designs which separate the design data more and more accurately presumably also will separate the test data more and more accurately. If it becomes apparent that the test data cannot be separated sufficiently well by a continuation of the iteration procedures, this first effort has failed. The many decisions, which we made before the first design effort began, will have to be reevaluated, and questions of the following soulsearching kind will have to be answered. Precisely what makes us think that members of Class A can be separated from members of Class B? Are the recorded data representative, and do they really contain the patterns in question? Is the separation surface we use of a suitable functional form or should some other form be used? Nothing is more valuable than a good hunch; have any hunches been overlooked, which could have been helpful in the modelling (Block n° 5)? Depending on the answers to such questions it may be decided to start all over from Step 1 or to consider the problem as being unsolvable with the state of the art. It would of course be helpful if one could estimate the complexity of the problem in advance, this seems however to be difficult in practice (Bongard 1970).

4. Factors that Influence the Design of a PRD

<u>4. 1 Factors Associated with the Pattern Classes.</u>

In the following Class n° $i, i = 1, 2, ..., N_c$, is referred to as C_i . With each class of patterns is associated a number of significant constants, the values of which may or may not be known to the designer. The constants are related to the following questions that may be asked about each class of patterns.

(a) What is the <u>a priori probability,</u> $P_{pr}(i)$, of an unlabelled pattern being a member of C_i , rather than being a member of one of the (N_c-1) other pattern classes (Jaynes 1968). It may usu ally be assumed as will be done in the following that the patterns will arrive according to some time-stationary first order prob ability, and that they are not "maliciously" arranged in any manner. The assumption is not always satisfied. Sometimes in the class membership of arriving patterns show a Markov -dependence (Raviv 1967 Hilborn and Lainiotis 1968 and 1969; Lainiotis 1970). At other times as in the case of letters in a text there are strong statistical relationships between adjacent letters (Thomas and Kassler 1967).

(b) What is the cost, K_{ij} , of classifying a member of C_i as being a member of C_j ? In the following it is assumed that K_{ij} does not change with the percentage of misclassified members of C_i ; utility theoretical concepts are not applied. The convention will be used that K_{ij} , $i \neq j$, is non negative and $K_{ij}, i=j,$ is non-positive.

(c) Do the patterns change with time? If so, the parameter val ues used in the PRD must be adjustable and the values must be updated regularly (Amari 1967; Figuereido 1968; Hilborn and Lainiotis 1968). Unless otherwise indicated it is assumed in the following that all distributions of pattern attribute-values are stationary in time.

(d) How are the members of the same class related to each oth er, what it the nature of the relationship? Do the members of each class consist of a typical pattern, a prototype, corrupted

by noise? Or are the members of each class rather a collection
of well defined, noisefree patterns, which have certain features
in common (e. g., perfectly noisefree letters may be drawn in
a great many ways (Akers and Rutter 1964))? Or do the class
members have certain features in common while they at the
same time are all somewhat corrupted by noise? Throughout
the report it is assumed that the situation is basically probabi
listic so that members of each class have certain features in
common while they at the same time are all somewhat corrupted
by noise.

(e) Is there reason to believe that C_i consists of several distinct
subclasses, $C_{i,1}, C_{i,2}$, etc? If so, the multivariate distribution of
attribute values for members of C_i may very well be multimodal
(meaning that the density of C_i-pattern-points has several local
maxima in p-dimensional pattern space) or even disjointed
(meaning that any path which connects two local maxima in pat
tern space must have at least one segment of finite length in a
region where the density of C_i-points is zero). Multimodal,
multivariate distributions can create difficult problems in the
categorization procedure as will be discussed in Article 7.2.2.

(f) Are the members of each class known to the designer through
"representative" members of the class? In such cases it is of
greatest importance to ascertain that the sample members were
representative when recorded, and that the processing in the
data collection phase was performed with care so that the mem-

bers remained representative. In the following it will be assumed
that patterns which are taken to be representative of members
of C_i are indeed representative of C_i members; with certain
kinds of patterns (e. g. medical records such as EKGs) labelling
(or classification) of the pattern cannot be done with 100% accuracy. In the case where the designer only has a representative
collection of unlabelled patterns, he can attack the design prob
lem by one of the available methods; a review of such design
methods can be found in Chien and Fu 1967, Part IV.

Sometimes the designer uses data which previ
ously have been used by other researchers for similar projects.
In such cases the designer is able to directly compare the performance of his PRD with the performance of the previously
constructed machines. Some years ago a set of handprinted
characters was made available for public use (Highleyman 1963;
Munson et al. 1968) and presently the IEEE Pattern Recognition
Subcommittee on Reference Data is compiling a readily available, well documented body of data for designing and evaluating
recognition systems.

4.2 Estimation of a PRD's Performance

It may be instructive first to consider the following two special cases: use of "no attributes" and use of a set
of "arbitrary attributes". No attributes is the limiting case where

the designer simply does not design a receptor. In this case the designer uses only a priori information in the categorizer. A reasonable procedure would be to classify all unlabelled patterns as belonging to the class C_j for which the average cost of classification A_j,

$$A_j = \left[\sum_{\substack{i=1 \\ i \neq j}}^{N_C} P_{pr}(i) \cdot K_{ij} \right] + P_{pr}(j) \cdot K_{jj}$$

is minimized. The figure $(A_j)_{Min}$ is of importance to potential users when the justifiable cost of a PRD with a specified perfor mance is determined. If the designer selects a set of N_H attrib utes without regard for their effectiveness (and thereby saves some design effort) he has obtained a set of arbitrary attributes. The mean accuracy which can be obtained with a PRD using arbitrary attributes has been studied and was found to be discouraging (Hughes 1968). The designer is almost always well advised to spend a large part of his design effort in selecting an effective set of attributes; the problem of finding an effective set of attributes will be discussed in Subsection 5.2 and 5.3.

In the remaining part of this subsection the usual case is considered where the designer generates $N_p, N_p \geq N_H \geq p$, attributes each of which there is reason to believe will be effective; from this set of attributes the designer selects a particular effective subset of p attributes by a suitable procedure such as the ones described in Subsection 5.6.

4.3 Four Major Problem Areas

In the remaining part of this paper, the four major problem areas in the design of PRDs are briefly reviewed. The problem areas are listed below.

(a) The selection of an effective set of attributes for the description of patterns in question. This is the problem of what to measure.

(b) The selection of a decision procedure. This is the problem of how to categorize the receptor output, Figure 1.

(c) When the decision procedure has been selected there is usually a set of parameters to be evaluated. Here the problem is encountered of how to optimize the parameter values with respect to given representative patterns from the N_C different pattern classes.

(d) The realization in hardware of the receptor and the categorizer. Here the problem is raised of selecting forms of receptors and categorizers which will perform reliably when implemented, and which will meet the PRD constraints on weight, volume, etc.

5. The Selection of the Attributes

The problem of how to select an effective set of attributes for a PRD is generally considered the single most difficult problem in the design. The problem is frequently discussed in the literature (Levine 1969; Nagy 1969; Nelson and Levy; Fu et al. 1970; Henderson and Lainiotis 1970) and is the topic of September 1971 issue of the IEEE Transactions on Computers.

This section is organized as follows. First the preliminary processing of patterns and the generation of attributes is discussed. In Subsection 5.3 the important concept of an "effective set of attributes" is defined. Next some one-number statistics are reviewed which may be used to measure how good one attribute is when used alone. After this, in Subsection 5.5 an important set of attributes, the templets of N_c prototypes, is described. The necessary concepts have now been developed to discuss ways of selecting an effective set of p attributes, this is being done in Subsection 5.6.

5.1. Preliminary Processing.

The main operation in the receptor, Figure 1, is the computation of p attribute values, $\Xi_1, \Xi_2, ..., \Xi_p$, for the unlabelled pattern. Before the q measurements, $\Upsilon_1, ..., \Upsilon_q$, can be performed it often happens that the pattern must be isolated because other patterns and the background tend to obscure the the pattern in question, e. g. letters in handwritten text or objects in reconnaissance photographs. Next the pattern may be exposed to preliminary processing where the designer takes advantage of his a priori knowledge of the recognition problem. E.g., (i) suitable filtering may be used to enhance the signal to noise ratio for waveforms and images (Pattern Recognition, vol. 2, n°2, May 1970 contains 7 papers on Image Enhancement, vol. 2, n°1, January contains 5 papers on two-dimensional Picture Processing), (ii) handwritten alpha-numerical characters may be normalized with respect to size and orientation (Casey 1970), (iii) irregularities that clearly are due to noise may be removed from the unlabelled pattern by "smoothing" so that edges become clearer (Unger 1959; Levine 1969; Golay 1969; Deutsch 1970) and (iv) the dynamic range for speech signals may be reduced (Hellwarth and Jones 1968).

As an example of preliminary processing consider infinite clipping of waveform patterns. A waveform $v = f(t)$ is said to be infinitely clipped when only the sign of the amplitude, signum $\{f(t)\} = \text{sgn}\{f(t)\}$, is retained and the magnitude, $|f(t)|$ is discarded. After infinite clipping the waveform has been re duced to a record of the zero crossing times. Infinite clipping may seem to cause a great loss of information. This impression is, however, not always correct. E.g., it has been conjectured,

(Good 1967) that no loss of information whatsoever is suffered when a white noise signal is infinitely clipped as long as the signal is band-limited to the range (W_1, W_2) and $W_2/W_1 < (7 + \sqrt{33})/4 = 3,186$. Even when some information is lost, it is not necessarily the informa tion of interest to the designer. E.g., it has been found that the intelligibility of a clipped speech signal remains high al- though the quality of the reproduced sound has become poor (Licklider 1950; Fawe 1966; Scarr 1968).

5.2. Generation of Sets of Attributes in Practice.

The designer can obtain some guidance in his search for a set of effective attributes by consulting with ex- perts in the area of interest, by studying the literature and by inspecting typical patterns. The possibilities are, however, often limited not only by restrictions on the PRD with regard to size, weight, cost and processing time per pattern but also by constraints particular for the problem at hand; two typical con straints will now be mentioned. (i) The designer may be re- stricted to attributes, the values of which he can measure with out storing the complete pattern (Bonner 1966); the frequencies of occurrence of specified short binary words in long binary sequences is an example of such attributes. (ii) In many pattern recognition problems the designer must look for attributes that show invariance under the commonly encountered forms of dis tortion; e. g. invariant attributes can be generated by moments (Hu 1962), autocorrelations (Mc Laughlin and Raviv 1968), integral geo-diffraction-pattern sampling (Lendaris and Stanley 1970), integral geometry (Tenery 1963) and by application of Lie group theory (Hoffmann 1965). Examples of transformation invariant attributes may be found in papers indicated by the descriptor D_{12} in Minsky 1963.

Often the patterns can be expanded using a set of orthonormal functions; in such cases the coefficients to the p most important terms of the expansion may be used as attrib utes. The designer will select a set of orthonormal functions which is simple to store in the receptor and which converges rapidly to the type of patterns in question; some of the possibilities are: (i) "The fast Fourier transform" (Cochran et al. 1967) which makes use of the Cooley-Tuckey algorithm, (ii) the Rademacher-Walsh functions (Harmuth 1968), Haar functions (Bremermann 1968, Section VIII) and related expansions for functions of binary variables (Ito 1968), (iii) the Karhunen-Loeve expansion (Chien and Fu 1967b), and (iv) Wiener's Hermite-Laguerre expansion (Brick 1968).

To generate an effective set of p attributes two things are usually needed: the p attributes should be selected as a best (or at least relatively good) subset from a larger set of attributes and some evolutionary procedure (as for instance the FOBW method described in Becker 1968 and in Appendix 4) should exist for improving the subset of p attributes. The selection of the attributes is probably the single most important step in the PR design; a witty description of the search for good attributes and some of the pitfalls has been given by Selfridge (Selfridge 1962).

5.3   An Effective Set of Attributes.

The problem in receptor design is to find an effective set of N_H or fewer attributes, where N_H is the largest number of attributes which the designer can afford to use in the later hardware (or software) implementation; p in Figure 1 is the number of attributes that is actually used, p is usually equal to N_H. In rare cases when fewer than N_H attributes give adequate performance, p is less than N_H. The material in this subsection is divided as follows. First the concept of "an effective set of p attributes" is defined in Article 5.3.1. This definition leads to a definition of "the incremental effectiveness of an attribute Ξ_{p+1}" presented in Article 5.3.2.

5.3.1.   A Definition of "An Effective Set of Attributes".

The following definition is based on a discussion of "effectiveness of receptors", which has appeared in the literature (Marill and Green 1963). Let it be assumed that the following are available to the designer.

(1) Two sets of p attributes S_1 and S_2 which may have several attributes in common; this is in principle equivalent to having

two receptors R1 and R2 specified.

(2)   A number of representative patterns that are divided into a Design Data group and Test Data group.

(3) A means for categorization. It could be a certain computer program for design of categorizers which are optimum in some sense, e.g., the number of misclassified patterns may be mini mized by a computed linear separation surface. One such com- puter program will be described in Appendix 2. Other means for categorization and the problem of designing categorizers will be discussed in Section 7.

(4)   An index of performance, IP, by which the performance of a PR can be evaluated; the convention will be used that higher values of IP illustrate better (rather than poorer) performance. The IP is largely determined by the decision procedure the de signer uses (decision procedures will be discussed in Section 6). An example of an IP is "minus one times the average cost per classification of a test data pattern"; this IP will be used by the designer in case he uses Bayes' decision. Notice that the de- signer in some practical cases may have to use a means for cat egorization that optimizes a feature (e.g., least number of mis classified patterns) that does not necessarily coincide with the IP (e.g., the average cost of classification, when $K_{ij} \neq K_{ji}$).

The designer now takes the following steps. (A) With the design data and R1 he computes the optimum categorizer, C1. He separates the test data using the PR consisting of R1 fol lowed by C1. He computes the value of the index of performance for the test data, the value is called IP(1). (B) With the same design data and R2 he computes the optimum categorizer C2. He separates the test data using the PR consisting of R2 followed by C2. He computes the value, IP(2), of the index of performance for the test data. If IP(1) exceeds IP(2), the attribute set S_1 is said to be a more effective set of p attributes than S_2. Usually the performance of the PR has been specified in advance meaning that the index should have at least a certain specified value, IPS. If IP(1) is close to or exceeds IPS, S_1 is said to be an effective set of p attributes.

The concept of effectiveness is conditioned on the available means for categorization and (what is usually less important) on the index of performance. If a set of p attributes is to be effective, it is necessary (1) that the members of each class tend to cluster at certain locations in the p-dimensional space, and (2) that no two clusters from the N_c sets of clusters coincide. What is less obvious but also important especially for large p-values is that (3) each of the N_c probability densities be essentially unimodal or at least not essentially disjointed so that N_c populations can be well separated by p-dimensional surfaces of simple form. When the last condition has not been met, it is doubtful whether the means for categorization will be capable of performing a satisfactory separation. This matter will be discussed further in Article 7.2.2.

5.3.2 A Definition of "The Incremental Effectiveness of an Attribute".

Consider the case where the designer has (i) a set of p attributes, S_p , (ii) a set of (p+1) attributes, S_{p+1} , obtained by adding the attribute Ξ_{p+1} to S_p, (iii) a means for categorization, and (iv) an index of performance, IP. The designer can now compute the two values of the index, IP(p) and IP(p+1), in the manner described earlier. The PR designed with (p+1) attributes will always perform at least as well as the one designed with p attributes. The quantity IP(p+1)-IP(p) is called the incremental effectiveness of Ξ_{p+1}. The incremental effectiveness is conditioned on S_p, IP and the means for categorization.

5.4 One Attribute.

The attribute is called Ξ_1, as in Fig. 1. In this subsection it is discussed how the designer may estimate the effectiveness of Ξ_1. It is assumed that the probability density functions of the Ξ_1-values for members of Class 1, members of Class 2, etc. have been obtained; in practice normalized histograms are often used as approximations to the N_c probability density functions. It should be noticed that a set of p attributes each of which can take a finite number of discrete values, v_i, i=1,...,p, in a sense may be considered as one attribute with a

number of discrete values equal to the product: $v_1 \cdot v_2 \cdot \ldots \cdot v_p$.
E.g., if Ξ_1 can take the values 2, 3 or 4, and Ξ_2 can take the
values -1 and -5 then the "compound attribute"(Ξ_1, Ξ_2)can take
the six discrete values (2, -1), (2, -5), (3, -1), (3, -5), (4, -1),
and (4, -5).

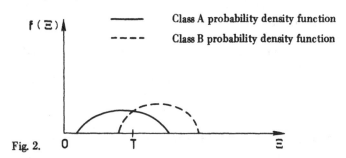

Fig. 2.

The members of Class A and Class B both with unimodal density functions may be separated by use of
one threshold, T.

   The following notation will be used frequently
in the remaining part of this report. The probability density
function for members of Class A is called $f_A = f_A(\Xi_1)$; f_A has the
mean a and the variance σ_A . The members of Class B have the
density function $f_B = f_B(\Xi_1)$ with mean b and variance σ_B . The functions f_A and f_B , and the moments a, b σ_A^2 and σ_B^2 , are usually
unknown to the designer. When it is stated in the following that
the designer uses f_A, f_B, a, b, σ_A , and σ_B , it usually means that
the designer uses estimates of f_A, f_B, a, b, σ_A , and σ_B obtained
from the normalized histograms.

   When Ξ_1 is an effective attribute it will be seen
that the N_c density functions overlap little or not at all. No matter what the functional form of a separation surface may be, in

one dimension categorization is achieved by establishing suita‐

ble Ξ_1-thresholds; Figures 2 and 3 illustrate two such cases.

The interval on the Ξ_1-axis in which the Ξ_1 -value of an un‐

labelled pattern falls decides the class to which the unlabel‐

led pattern is assumed to belong. This type of categorizer is

very simple to realize in hardware. When the designer sets the

threshold he will try to optimize some measure of separability;

"measure of separability" will in the following be abbreviated

MS will then give an estimate of how effective Ξ_1 is compared to

other attributes. Three possible MSs will now be listed.

(1) The average cost of classifying a pattern, Bayes' criterion

(Middleton 1960, Art. 1.8.4-3). This MS will be discussed fur‐

ther in Subsection 6.2.

(2) The minimaxed average cost of classifying a pattern. The

case $N_C=2$ has been discussed in the literature (Selin 1965, Chap‐

ter 2). The case $N_C > 2$ will be discussed briefly in Subsection 6.3.

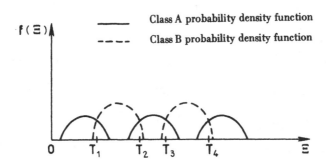

Fig. 3.
The members of Class A that have a trimodal, disjointed density function may be separated from the
members of Class B that have a bimodal, disjointed density function by use of four thresholds, T_1, T_2,
T_3, and T_4.

(3) Minimized error of the first kind for fixed value of error of the second kind, the Neyman-Pearson criterion; the MS can be used only when $N_C = 2$ (Davenport and Root 1958, Section 14-3; Selin 1965, Chapter 2), it will be discussed in Subsection 6.5.

These three MSs require that the designer has selected an optimum setting of the threshold(s); in the following the convention will be used that such MSs also are called indices of performance, IPs. IPs will be discussed in Section 6. There are, however, other MSs, the values of which may be computed directly from the distributions without a specification of the threshold settings. Four such MSs will now be listed.

(4) The average information, G_1, about class membership given by Ξ_1 (Lewis 1962; Kamentsky and Liu 1964). Let it be assumed that Ξ_1 can take v discrete values $\xi_1, .., \xi_l, .., \xi_v$, if so, G_1 is defined as follows (Lewis 1962).

$$G_1 = \sum_{i=1}^{N_C} \sum_{l=1}^{v} P(C_i, \xi_l) \log_2 \left[P(\xi_l | C_i) / P(\xi_l) \right]$$

A numerical example will illustrate the computation of G_1. Let N_C equal two and v equal three; and let

$$P(C_1, \xi_1) = .31 \qquad P(C_1, \xi_2) = .16 \qquad P(C_1, \xi_3) = .03$$

$$P(C_2, \xi_1) = .01 \qquad P(C_2, \xi_2) = .17 \qquad P(C_2, \xi_3) = .32$$

G_1 then becomes the sum of the 2.3 terms:

$$.31 \, \log_2\left[(.31/.50)/.32\right] + .16 \, \log_2\left[(.16/.50)/.33\right] + .03 \, \log_2\left[(.03/.50)/.35\right] +$$

$$+ .01 \, \log_2\left[(.01/.50)/.32\right] + .17 \, \log_2\left[(.17/.50)/.33\right] + .32 \, \log_2\left[(.32/.50)/.35\right]$$

$$G_1 \approx .52 \text{ bits}$$

(5) The symmetric divergence (Kullback 1959, page 6; Marill and Green 1963); this MS is applicable only in the case $N_c = 2$. The symmetric divergence J_{AB} is defined by the following expression

$$J_{AB} = \int_{-\infty}^{\infty} \left[f_A(\Xi_1) - f_B(\Xi_1)\right] \ln\left[f_A(\Xi_1) \, f_B(\Xi_1)\right] d\Xi_1 \,.$$

In the case with the six discrete probability listed above J_{AB} becomes the sum of three terms:

$$(.62 - .02) \, \ln(.62/.02) + (.32 - .34) \, \ln(.32/.34) + (.06 - .64) \, \ln(.06/.64)$$

$$J_{AB} \approx 3.4$$

(6) The Fisher measure $(a-b)^2/(\sigma_A^2 + \sigma_B^2)$; this MS was proposed by R. A. Fisher (Fisher 1963) and explored by Sebestyen (Sebestyen 1962). The MS is applicable only in the case $N_c = 2$.

(7) The measure of separability S which recently has been proposed (Becker 1968).

$$S = |a-b| / (\sigma_A + \sigma_B) \,.$$

This MS is applicable only in the case $N_c = 2$.

The designer must decide which of the seven MSs is most meaningful in this case; maybe he would prefer some other MS, the possibilities are by no means exhausted. When the designer has made his choice he can estimate the effectiveness of all the attributes of interest.

It will be shown in Subsection 7.4 that the problem of separating N_c pattern classes may be reduced to a set of problems of separating two classes. The constraint $N_c = 2$ is consequently not very restrictive in practice.

5.5 Templet Matching

Whenever the members of a class consist of a a class prototype plus some corrupting noise it is reasonable to use the agreements with templets of the N_c prototypes as attributes. The match between an unlabelled pattern and each of the templets is determined in the receptor. Next it is decided in the categorizer if the unlabelled pattern agrees well with one and only one of the templets. If so, the pattern is regarded as belonging to the corresponding pattern class; otherwise the pattern is rejected. The reading of printed or typed characters is typical of cases where templet matching can be applied with little or no preliminary processing of the raw data (Yau and Yang 1966). In other cases the templet is applied to a pattern obtained from the unlabelled pattern by a suitable transformation with some normalizing properties; e. g., the transformation of

of the unlabelled pattern could make use of integral geometry

(Tenery 1963) or the auto-correlation function (Horwitz and

Shelton 1961). A common problem with two-dimensional pat-

terns is to obtain a templet which is invariant to (i) vertical and

horizontal shifting, and (ii) rotation and magnification. One pos

sible way of solving the problem is to (a) generate the two-dimen

sional auto-correlation function (it is invariant to the shifting),

(b) plot the polar-coordinates of the auto-correlation function

in a cartesian coordinate system so that x equals the logarithm

of the radius vector and y the vectorial angle, (c) find the two-

dimensional auto-correlation function for the figure in the car

tesian system, it will be invariant to shifts in the vectorial angle

and to shifts in the logarithm of the radius vector; in other words

the templet is invariant to shifting, rotation and magnification.

            To use the templet matching technique it is

necessary to introduce some measure of the match between the

unlabelled pattern and each of the N_c templets. Papers that dis

cuss matching criteria are indicated by the descriptors D_2 and

D_3 in Minsky 1963.

            If it is known that only one templet will match

the unlabelled pattern the search can be stopped when such a

templet has been found. On the average $N_c/2$ templets must be

checked before the pattern is identified. When an exhaustive

search through all N_c templets is not required, two things can be done to accelerate the recognition procedure. (1) If the pat

terns appear in certain contexts (which is contrary to the as-

sumption in Paragraph (a) in Subsection 4.1) the templets may

be searched in such a manner that the most likely class member

ships (judging from the last few recognized patterns) are check

ed first. This situation is for instance encountered when the pat

terns are letters in an ordinary English text (Thomas and Kassler

1967). (2) If the number of classes is large, the classes may be

divided into a smaller number of super-classes, for each of

which a templet is developed. As an example: let the number

of super-classes be $\sqrt{N_c}$ and let each super-class contain $\sqrt{N_c}$

pattern classes. The templet matching may then be performed

in two steps. First, the correct super-class is located. On the

average, when there is no context information available $\sqrt{N_c}/2$

super-class templets must be checked before the super-class

is identified. Next, the templets for the members of the super-

class in question are checked. When there is no context infor-

mation available, $\sqrt{N_c}/2$ templets on the average must be check

ed before the class-membership is established. In the case of the

numerical example mentioned above, the two-step procedure

would on the average require a total of only $\sqrt{N_c}$ templet match

ing operations per classification. For N_c large, this is substan

tially less than the $N_c/2$ operations required with a one-step

procedure. A two-step procedure has for instance been used in

a device to recognize 1000 printed Chinese ideographs (Casey and Nagy 1966).

5.6 Selection of a Set of p Attributes.

In this subsection the general case is consider ed where the number of attributes to be selected, p, is more than one, and where the attributes are of a more general nature than the measurement of agreement between a pattern and a set of N_c templets. The problem of concern is: how should the designer in practice select an effective subset of p attributes from a large set of N_p attributes? The effectiveness of a set of attributes was defined in Article 5.2.1 and it was pointed out that the effectiveness is conditioned on (i) an index of performance, IP, related to the decision procedure used by the designer, and on (ii) "the means for categorization" meaning the device by which the decision space is partitioned. In practice it is usually not too important for the effectiveness which IP is used. The choice of the means for categorization, on the other hand, is crucial for the effectiveness of a set of attributes; e. g., the attribute pair(Ξ_1, Ξ_2)in Figure 4 will give perfect separation if and only if the means for categorization can establish a separation surface consisting of two vertical and two horizontal lines. In the remaining part of this subsection it will be assumed that the de signer has selected an IP and a means for categorization which are suited for the problem at hand.

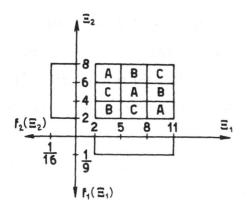

Fig. 4. A Bivariate Distribution.

The members of Class A are located in the three 3 x 2 rectangles marked A ; the tri-modal, bi-variate probability density is constant and equal to 1/18. The members of Classes B and C like-wise are located in the rectangles marked B and C, their density is also 1/18. The three classes of patterns may be perfectly separated if and only if the separation surface consists (essential-ly) of four straight lines, $\Xi_1 = 5$, $\Xi_1 = 8$, $\Xi_2 = 4$ and $\Xi_2 = 6$. If the members of all three classes had been distributed each with the density 1/54 over the rectangle with the vertices (2,2), (2,8), (11,8) and (11,2), the classes would have been inseparable with any kind of separation sur-face. Yet, in both cases the marginal densities $f_1(\Xi_1)$ and $f_2(\Xi_2)$ would be the same for all three classes. This example illustrates the point that even when the marginal distributions for the different classes do overlap much, perfect separation may be possible when the separation surface is compatible with the multivariate distributions.

In cases where the designer does not want to commit himself to an IP and a means for categorization he may estimate the ranking of the effectiveness of subsets of p attri-butes through use of an indicator which is independent of the location of the separation surface. The designer could, for in stance, use (i) the upper bound of an error probability (Yau and Lin 1968), (ii) the information measure G_1 (Lewis 1962; Kamentsky and Liu 1964), or (iii) the symmetric divergence, J_{AB}; G_1 and J_{AB} were mentioned in Subsection 5.4, paragraphs 4 and 5; in particular J_{AB} has been widely used. Notice that when a set of p attributes is used to separate N_C pattern classes the

the separation of any two classes can never be inferior to the

separation which could have been achieved with any subset of

of the p attributes. Next three different procedures for select-

ing an effective subset of attributes from a set of N_p attributes

will be described; a fourth approach (Akers 1965) will be describ

ed in Appendix 2.

Exhaustive Search. If the number of possible

subsets, $(N_p!/(p!\ N_p\text{-}p)!)$, is not too large the value of the IP may

be computed in each case. Of all the possible subsets the subset

with the largest IP-value is selected and this truly optimum sub

set is used. For $N_c > 2$ and normal distributions with equal cova-

riance matrices a procedure has been used (Chen 1965) where

the set of p attributes is so chosen that the minimum symmetric

divergence among the $(N_c\text{-}1)N_c/2$ pairs of classes is maximized;

the purpose of this selection procedure is to obtain a safeguard

against the choice of attribute sets that may result in large mis

recognition percentages for certain pattern classes. Usually,

however, the number of possible subsets is by far too large to

permit an exhaustive search.

Sequential Elimination. The IP-value is com-

puted for the $(N_p\text{-}1)$ possible subsets of $(N_p\text{-}1)$ attributes. The

subset with the highest IP-attribute is retained; let it contain

the attributes $\Xi_1, \ldots, \Xi_{N_p\text{-}1}$, the attribute $\Xi_{N_p\text{-}1}$, which had the small

est incremental effectiveness has thus been eliminated. Next

the IP is computed for the $(N_p\text{-}2)$ subsets obtained by suppress ing one of the retained attributes at a time. The attribute with the smallest incremental effectiveness, say Ξ_{N_p-1} is eliminated, etc. The procedure is continued until only p attributes remain; the set of p attributes is then regarded as being if not the optimum at least a near optimum subset of p attributes. A sequential elimination requires fewer computations than an exhaustive search and experimental evidence (Cardillo and Fu 1967, Paragraphs 1 and 2) tends to indicate that approximately the same subset of p attributes will be obtained by either method. The sequential elimination method has been used with the symmetric divergence, J_{AB} , as estimator for effectiveness and under assumption of normal distribution and $N_c = 2$ (Marill and Green 1963, Experiment 1).

The Matrix Method. In most practical cases both an exhaustive search and a sequential elimination procedure will require far too much computation effort. Usually the designer has available only a set of N_p density functions (or normalized hisotgrams) for each of the N_c pattern classes. The designer may now proceed as follows:

(i) draw a matrix of size N_c by N_c ; the entry at row n°i and column n°j is called A_{ij} ; only the $N_c(N_c-1)/2$ entries in the upper will be used;

(ii) list at each entry the names of those (among the N_p) attributes which are effective in separating members of C_i from members of C_j ;

(iii) select the p attributes one by one beginning with the attrib ute mentioned most often in the matrix and continue in a man ner which agrees with the index of performance (the designer had only the marginal distributions to start with so a heuristic selection procedure must be excused at this point).

Use of the matrix method has been described in the literature; as an example a model for pattern recognition by Kamentsky and Liu (Kamentsky and Liu 1964) will be briefly reviewed. It is here assumed that all attributes are two-valued functions and $N_c \geqslant 2$. Two indicators, I and D, are used to quanti tatively describe each attribute. I is the average information about class membership given by the attribute just like G_1 (Lewis 1962). When a set of attributes is used there is the possibility that the percentage of misclassification becomes very large for the members of one of the classes. To overcome this problem the concept of the quantity D is introduced. D_{ij} is defined as the number of attributes that separate members of C_i and C_j "well" in some appropriate sense. With this method the set of p (or fewer) attributes is selected so that the $N_c(N_c-1)/2$ D-val ues all exceed a certain specified minimum; the designer will try to select the p attributes in a manner so that the smallest D_{ij} becomes as large as possible.

6. Decision Procedures and Indices of Performance

Figure 1 illustrates that the receptor maps a pattern into a point in p-dimensional space. Based on the location of the point a decision is made in the categorizer with regard to which of the N_c pattern classes, C_i, $i=1,...,N_c$, the pattern presumably belongs to Ho and Agrawala 1968; Das 1969). The designer makes the decision so that a suitable IP is optimized for the design data, the value of the optimized IP indicates the relative performance of the PR.

This subsection describes four methods by which such a decision may be made. The methods are Bayes' procedure, the minimaxing classification procedure, the likelihood method and the Neyman-Pearson method. The result of any of the four decision procedures may be realized through explicit partitioning of the pattern space. The result of Bayes' procedure and the likelihood method may also be realized by implicit partitioning (Minsky 1961, Section G-1). The four decision procedures mentioned above are well known and widely used. The designer is of course free to define a new IP, say a "mean-squared error"-criterion (Amari 1967), if that would be more appropriate for the problem at hand.

6.1 Some Functions Related to the Micro-Regions.

The problem of finding a suitable decision pro cedure arises in the following manner. In this section assume that such a large number of representative members of all N_C pattern classes are available that the N_C multivariate probabil ity densities can be estimated with great accuracy over the whole p-dimensional space. This assumption is unrealistic in most cases; e.g. if 8 attributes each can take 10 discrete val ues it becomes necessary to estimate N_C probabilities at 10^8 points. In practice the problem may be handled by one of the methods described in Subsection 7.1. Let the space be partition ed into a very large number, N_M, of very small p-dimensional compartments, the micro-regions, $W_1, W_2, \ldots, W_k, \ldots, W_{N_M}$. A point that represents a pattern is assumed to fall in one and only one of the micro-regions; such a pattern will be called Ξ. If a point represents a member of C_i, it is referred to as a C_i-point and it will be called $_i\Xi$. Let the probability with which a C_i-point falls in micro-region W be $P(i)_k$.

(1) $$P(i)_k = P\left(\Xi \in W_k \mid \Xi \text{ is a } _i\Xi\right)$$

Let $P_{pr}(i)$ be the a priori probability of an unlabelled pattern belonging to pattern class C_i. The probability of an unlabelled pattern being represented by a point that falls in micro-region W_k is then P_k.

$$P_k = \sum_{m=1}^{N_C} P(m)_k \cdot P_{pr}(m). \qquad (2)$$

The sum of the N_M P_k -values, $k = 1, \ldots, N_M$ is unity. Let K_{ij} be the cost of classifying a member of C_i as being a member of C_j . The convention is used that for any given i, K_{ii} is non-positive and the (N_C-1) K_{ij} -values, where $i \neq j$, all are non-negative. To describe the cost of all possible correct and er-roneous classifications, a matrix with N_C by N_C entries is needed. When an unlabelled pattern is represented by a point that falls in the micro-region W_k , the pattern will be classified as belonging to one of the N_C classes, say C_i . Sometimes this decision is correct and the cost K_{ii} is incurred. If the pattern belongs to one of the other classes a cost K_{ji} is incurred, $i \neq j$ On the average the cost $(K_i)_k$ is incurred if it is decided that all patterns that are represented by points falling in W_k should be classified as members of C_i .

$$(K_i)_k = \sum_{m=1}^{N_C} P(m)_k \cdot P_{pr}(m) \cdot K_{mi} \qquad (3)$$

$(K_i)_k$ may also be expressed by Equation 4.

$$(K_i)_k = E\left\{ \text{classification Cost} \mid \Xi \in W_k, \text{all } \Xi \text{ classified as } {}_i\Xi \right\} \cdot P_k \qquad (4)$$

6.2 Bayes' Procedure

With each class C_i and each micro-region W_k is associated the expected cost of classification $(K_i)_k$. An eminently reasonable rule for classifying an unlabelled pattern that is represented by a point in W_k is to classify it as belonging to the pattern class $i(k)$, for which the average classification cost is as low as possible;

$$\left(K_{i(k)}\right)_k = \min\left\{(K_i)_k\right\}, \quad i \in \left\{1, \dots, N_C\right\}.$$

This procedure of classifying so as to minimize the expected cost was developed by Thomas Bayes, about two hundred years ago. With this procedure an expected cost R^* is associated with the classification of an unlabelled pattern; R^* can be computed from equation 6, which may be derived in the following manner. Some members of C_j will be classified correctly, namely those for which the corresponding points fall in micro-regions where all points are classified as C_j -points; such micro-regions have $i(k) = j$. The remaining numbers of C_j will be misclassified in one or more of the (N_C-1) possible ways. Equation 5 determines the expected cost, $R(j)$, of classifying a pattern, given that the pattern belongs to C_j . $R(j)$ should not be confused with $(K_i)_k$ which was defined by Equation 3. N_M is the number of micro regions.

$$R(\dot{\iota}) = \sum_{k=1}^{N_M} P(\dot{\iota})_k \cdot K_{\dot{\iota}\dot{\iota}(k)} \cdot \qquad (5)$$

It is now clear that "Bayes' risk", R^*, the minimized, expected cost of classifying an unlabelled pattern, must be the weighted average of $R(\dot{\iota})$ as indicated by Equation 6.

$$R^* = \sum_{\dot{\iota}=1}^{N_C} R(\dot{\iota}) \cdot P_{pr}(\dot{\iota}) \cdot \qquad (6)$$

When Bayes' procedure is used "minus one times the average cost of classifying a pattern" would constitute a reasonable index of performance.

Usually not all $R(\iota)$, $\iota \in \{1, \ldots, N_c\}$; have the same value. Let it be assumed that the following relationship exists:

$$R(1) \geqslant R(2) \geqslant \ldots \geqslant R(N_c).$$

Recalling Equation 6 the following relationship is seen to be valid.

$$R(1) \geqslant R^* \geqslant R(N_c).$$

6.3 The Minimaxing Classification Procedure.

If the a priori probabilities for some reason change from the given values, $P_{pr}(\iota)$, the designer must rede sign the categorizer, meaning that he must find new values for $i(k), k=1,\ldots,N_M$. In case the classification of points in each micro-region has to be maintained by the designer after the change of a priori probabilities, the average classification cost presumably will change. Assuming that $R(1) \geqslant R^* \geqslant R(N_c)$ the cost could increase to $R(1)$ if all unlabelled patterns now belong to C_1, and it could decrease to $R(N_c)$ if all unlabelled patterns belong to C_{N_c}. If the designer has no information about the values of $P_{pr}(1),\ldots,P_{pr}(N_c)$ he cannot afford to guess; he will want to classify in such a manner that the largest possible average classification cost has been minimized. To do this the designer will make use of the minimaxing classification procedure, a procedure that will now be described in principle. All patterns that are represented by points falling in W_k will be classified as belonging to C_{i_k}, $i_k \in \{1,2,\ldots,N_c\}$. i_k will be defined shortly; i_k should not be confused with the minimizing quantity $i(k)$ from Bayes' procedure. Any classification rule may be described by an array of N_M i_k-values; let $(i_1,\ldots,i_{N_M})$ illustrate the array. The average cost of classifying members of C_j is $B(j), j=1,\ldots,N_c; B(j)$ is defined by Equation 7 for the minimaxing classification procedure. $B(j)$ may be considered a counterpart to $R(j)$, Equation 5.

(7)
$$B(j) = \sum_{k=1}^{N_M} P(j)_k \cdot K_{j\,i_k}.$$

The minimaxing procedure consists of finding the array $(i_1,\ldots,i_{N_M})$ for which the maximum $\{B(j)\}$ is minimized, $j \in \{1,\ldots,N_c\}$.

$$\text{minimax}\left\{B\left(\dot{\imath}\right)\right\} = \underset{(i_1,\ldots,i_{N_M})}{\text{minimum}}\left\{\text{maximum}\left\{\sum_{k=1}^{N_M} P(\dot{\imath})_k \cdot K_{\dot{\imath}i_k}\right\}\right\} \qquad (8)$$

Equation 8 shows how $\text{minimax}\left\{B(\dot{\imath})\right\}$ in principle may be found by (i) changing one of the N_M i_k -values at a time and by (ii) always making the change in such a manner that the largest of the N_c average clsssification costs is reduced. In the general case there may be several, local minimax values of different magnitude. When $N_c = 2$, the result of the minimaxing procedure may be expressed in closed form (Selin 1965, Chapter 2). It is possible to mix the Bayes' and the minimaxing decision procedures so that one minimaxes to avoid catastrophically high risks and minimizes the remaining average risk (Lehman 1959, Section 1.6).

When the minimaxing procedure is used (-maximum$\left\{B(\dot{\imath})\right\}$) would constitute a reasonable index of performance.

6.4 The Likelihood Method.

In many cases the values of the N_c^2 possible classification costs, and the N_c a priori probabilities are unknown to the designer. In such situation only the N_c of $P(m)_k$ are available; $P(m)_k$ is the probability of a C_m -point falling in the micro-region W_k rather than in one of the (N_M-1) other micro regions, $P(m)_k$ was defined by Equation 1. A more or less obvious basis for a decision rule is given by what is called the likelihood principle. The principle states that an unlabelled pattern that is represented by a point located in W_k should be classified as being a member of the class C_{ℓ_k} for which class membership has the greatest likelihood; ℓ_k is defined by Equation 9.

$$(9) \qquad\qquad P\left(\ell_k\right)_k = \underset{m}{\text{maximum}}\left\{P(m)_k\right\}, m \in \left\{1,...,N_c\right\}$$

If class n° s happens to be C_{ℓ_k}, the (N_c-1) likelihood ratios $P(s)_k /$ $/P(t)_k, t=1,...,N_c, t \neq s$, will all be somewhat larger than unity. If the (N_c-1) likelihood ratios are computed for any of the other (N_c-1) classes at least one of the likelihood ratios will be less than unity. It may consequently be determined from the likelihood ratios which class has the highest likelihood of class membership. Algorithms for approximating likelihood ratio computations have been described in the literature (Sebestyen 1962 Section 4.3). The likelihood method and Bayes' procedure become identical in the often encountered case where (i) all N_c a priori probabilities are equal, $P_{pr}(i) = 1/N_c$, (ii) all N_c correct classification costs K_{ii} are equal, and (iii) all $N_c(N_c-1)/2$ misclassification costs $K_{ij}, i=j$, are equal.

When the likelihood method is used the percentage of correctly classified patterns would be a reasonable index of performance.

6.5 The Neyman-Pearson Method.

When as before the a priori probabilities and the classification costs are not known to the designer and furthermore $N_C = 2$, a special procedure becomes possible. Classification of points in the N_M micro-regions may then be made according to the Neyman-Pearson criterion where the idea is to keep the percentage of misclassified C_1-patterns at some prescribed level, $100 \cdot R_{N-P}\%, 0 \leqslant R_{N-P} \leqslant 1$, while the percentage of misclassified C_2-patterns is minimized (Davenport and Root 1958, Section 14-3). The procedure is used when misclassification of C_1-patterns is particularly undesirable. The decision procedure will in practice consist of the following steps. (The procedure is not restricted to the familiar case, $p = 1$).

(i) First the likelihood ratio $L_k = P(2)_k / P(1)_k, k = 1, \ldots, N_M$, is computed for each of the N_M micro-regions.

(ii) The micro-region for which L_k is smallest, let it be W_1, is classified as a C_1 region and it is checked that the following inequality holds: $1 - P(1)_1 \geqslant R_{N-P}$.

(iii) Of the remaining micro-regions, the one with the by now lowest value of L_k, is classified as a C_1 region, let it be W_2. It is checked that $1 - P(1)_1 - P(1)_2 \geqslant R_{N-P}$.

(iv) In this manner repeatedly the micro-region with the lowest L_k-value is classified as a C_1 region. At each step it is checked that the percentage of C_1-patterns that thus far has not been classified does exceed $100 \cdot R_{N-P}\%$.

(v) Sooner or later micro-regions containing $100 \cdot (1-R_{N-P})\%$ of the C_1-patterns have been classified as C_1-regions. The remaining micro-regions are then all classified as C_2-regions.

By this procedure the N_M micro-regions have divided so that while $100 \cdot R_{N-P}\%$ of the C_1-patterns are misclassified the least number of C_2-patterns are misclassified.

When the Neyman-Pearson method is used and R_{N-P} is specified, the percentage of correctly classified C_2-patterns would constitute a reasonable index of performance.

6.6 Three Practical Difficulties.

What has been obtained this far in this section is development of some techniques by which all unlabelled patterns, which are represented by points in the micro-region W_k, may be classified as being members of one and only one of the N_C possible pattern classes. With these techniques so to say a class membership has been assigned to each micro-region. The next step in the development of a realistic decision procedure is the fusion of all adjacent micro-regions to which the same class, C_i has been assigned. By repeated fusing of the micro-regions a few large regions, macro-regions will be obtained; with each macro-region is associated a class membership. In case (i) a description of the p-dimensional surface that separates the macro-regions and (ii) the class number, i, assigned to each macro-region could be stored in the categor<u>izer</u>, a categorizer would have been obtained that was optimum in the sense chosen by the designer. In practice the approach described above - the construction of an "optimal recognition function" (Marill and Green 1960) - seldomly works out for three reasons.

(1) There are usually too few representative patterns available to determine the N_C multivariate distributions (without making assumptions about the distributions, a possibility which has not been considered in this section but will be discussed in Sub<u>section 7.1).</u>

(2) There is usually not enough time and funding to process the many patterns even if they were available.

(3) Unless the separation surfaces are of fairly simple shape or the dimensionality p, of the space is very low, the number of constants needed to describe the surfaces easily becomes so large that the information storage in the categorizer presents problems; in the extreme case where the micro-regions do not fuse at all it becomes necessary to store the class membership for each of the N_M micro-regions.

          In Section 7 it will be described how categorizers are designed in practice or in other words, how the p-dimensional separation surface, or an approximation hereof, may be determined in spite of the three obstacles listed above.

7. Categorizer Design

The answer to the three problems described in Subsection 6.6 clearly is, to partition the pattern space in a manner which requires the determination of fairly few constants. The few constants usually can be evaluated from the available representative patterns (which in most cases are limited in number) with a reasonable effort in time and funding, and the values of the few constants can be stored in memories of realistic size. This section about categorizer design is divided into four parts, one about methods for explicit partitioning of the pattern space, Subsection 7.2, and one about methods for implicit partitioning of the pattern space, Subsection 7.3. In both parts some methods the so called non-parametric methods, will be mentioned where the designer makes no assumption about the distribution in pattern space of the members of the N_c classes. With the remaining methods, the parametric methods, the designer in some manner makes use of known or assumed distributions. In Subsection 7.4 several methods are described for categorization of members from more than two classes, $N_c > 2$. As a preliminary it is in Subsection 7.1 considered how the N_c multivariate density functions may be estimated.

7.1 Estimation of a Multivariate Density Function.

By using p attributes rather than one in the receptor, the descriptive power of the receptor is enhanced substantially. However, at the same time a new problem has been created in the following manner. Whereas one-dimensional histograms usually can be obtained in practice, p-dimensional histograms are not available in most realistic situations be-cause their computation and subsequent storage become imprac tical. If **p**, for instance, equals 10 and each attribute can take 8 values, there are 8^{10}, or more than one billion, possible com binations of attribute values. It is usually not possible in prac-tice to estimate the multivariate density functions for each of the N_C classes with any precision over such a space; (for moder-ate values of p it may still be possible to obtain reasonable p-dimensional histograms (Sebestyen and Edie 1966). It con-sequently becomes necessary to make assumptions about the form of the density functions in some manner. When a suitable assumption can be introduced only relatively few constants have to be estimated before the density function can be estimated. Some examples will now be presented of the kinds of assum-ptions which are used by designers. The multivariate density function for members of C_i is called $P_i(\Xi) = P_i(\Xi_1, \Xi_2, ..., \Xi_p)$. The univariate density function for values of attribute Ξ_i is cal-led ${}^i P_i(\Xi_i), i=1,...,p$; the designer usually has enough patterns to

estimate these p marginal densities with sufficient accuracy
for each of the N_c classes.

Certain Higher Order Statistical Dependencies
are Disregarded. An often used (and strong) assumption is that
of statistical independence for all p attributes; in this very simple case $P_i(\Xi)$ becomes the product of the p (reasonably well
defined), marginal densities.

$$(10) \qquad P_i(\Xi) = {}^1P_i(\Xi_1) \cdot {}^2P_i(\Xi_2) \cdot \ldots \cdot {}^pP_i(\Xi_p)$$

Sometimes the designer for physical reasons can assume that
certain groups of attributes are statistically independent (this
is a weaker assumption than the one above); e.g., a first group
could be the first 3 attributes and a second group the remaining
(p-3 attributes. In such a case $P_i(\Xi)$ can be written in the following simpler form.

$$(11) \qquad P_i(\Xi) = {}^\alpha P_i(\Xi_1, \Xi_2, \Xi_3) \cdot {}^\beta P_i(\Xi_4, \ldots, \Xi_p)$$

A numerical example will illustrate the advantage of the decomposition. Assume that (i) all attributes are
three valued, (ii) $p = 6$, (iii) $N_c = 2$, and (iv) the number of
design data patterns from Class 1 and 2 is $M_{1D} = M_{2D} = 3^6 = 729$. If
so, the designer can estimate ${}^\alpha P_1, {}^\beta P_1, {}^\alpha P_2$ and ${}^\beta P_2$; each of the
four densities will consist of $3^3 = 27$ discrete probabilities which
may be estimated reasonably well given 729 pattern points. If
the designer does not make use of the decomposition he will

have to estimate P_1 and P_2 directly, P_1 and P_2 do both consist

of $3^6 = 729$ discrete probabilities; the 729 (rather than 27) esti

mates are still based on only 729 pattern points. It should now

be clear that the product of the $^\alpha P_1$ - and $^\beta P_2$ -estimates will be

at least as accurate and probably much more accurate than a

direct P_1 -estimate; also the product of the $^\alpha P_2$ - and $^\beta P_2$ -esti

mates can be expected to be substantially more accurate than

a direct P_2 -estimate. The more the designer can decompose

P_i , the easier and more accurate the estimation of P_i be-

comes.

The case where all p attributes are two val-

ued functions has received particular attention; techniques are

available for the approximation of $P_i(\Xi)$ to reduce storage re-

quirements (Lewis 1959), $P_i(\Xi)$ may for instance be approxi-

mated with dependence trees (Chow and Liu 1968; Ku and Kull

back 1969) or Markow chains (Chow 1966).

A Specific Functional Form is Assumed. Some

times the designer knows (or guesses) what the functional form

$P_i(\Xi)$ is. The functional form could be Gaussian, Pearson Type

II or Type VII (Cooper 1964), uniform density inside a hyper-

sphere, etc. In such cases the designer usually pools the design

data for each class and estimates the values of the parameters

of interest, e. g., for a Gaussian distribution the important para

meters are the mean vector and the covariance matrix for each

class. In rare cases the designer in addition has statistical

information about the parameter values he is about to estimate
and becomes able to estimate the values in a sequential fashion.
E.g., the designer may know a priori that the density is Gaussian with zero mean and that the inverse of the (unknown) covariance matrix has a Wishart density (Keehn 1965, Part III).
Having measured the attribute values for the first representative
pattern he is able to compute the a posteriori probability density
for the inverse covariance matrix, the density turns out also
to have the form of a Wishart density (but it is more concentrated than the previous one). In such special cases where the
a priori and a posteriori densities for the unknown parameters
have the same functional form (which is the exception, see Lin
and Yau 1967, Table 1), it becomes possible to estimate the
parameter values by repeating the procedure; each member of
C_i in the design data group is used to obtain a new and more
concentrated density for the unknown parameter values. When
all the representative C_i members have been used, the expected values of the parameters are used as true values in the $P_i(\Xi)$
-density function.

Truncated Orthonormal Expansion. Based on
his knowledge (or expectations) about $P_i(\Xi)$, the designer selects a suitable orthonormal system, $\Phi_1(\Xi), \Phi(\Xi)$, etc., and decides on how many terms, say n, he will use for his approximation of $P_i(\Xi)$. The remaining questions are: what weights w_1,
$w_2, \ldots, w_n$ should the designer use to make $\sum_{i=1}^{n} w_i \Phi_i(\Xi)$ a best

approximation of $P_j(\Xi)$ in the mean square error sense, and what is the rate of convergence (Wagner 1968)? The answers may be obtained by stochastic approximation (Robbins and Monro 1951; Kashyap and Blaydon 1966; Laski 1968) or related techniques (Kashyap and Blaydon 1968), where the weight vector $w = (w_1, \ldots, w_n)$ is determined by an iterative procedure; also in the case of a non-linearly envolving non-stationary environment such techniques have been useful (de Figueiredo 1968; Yau and Schumpert 1968).

7.2 Explicit Partitioning of the Pattern Space.

This subsection is divided as follows. First some widely used separation surfaces of simple shape are mentioned, after which the need for such simple surfaces is established. Finally in Articles 7.2.3 and 7.2.4 parametric and non-parametric methods are described for locating such simple surfaces in an optimum manner.

7.2.1 Separation Surface of Simple Shape.

When the partitioning is explicit, use is made of separation surfaces of simple shape, e.g., quadrics (which illustrate second order equations), orthotopes (which are hyperspace-analogs of three dimensional boxes), and above all hyper

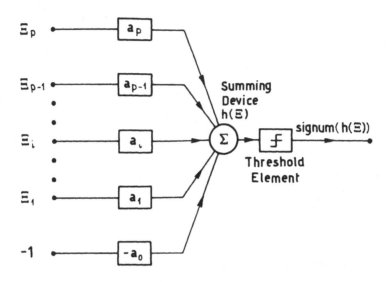

Fig. 5.

The threshold logic unit, TLU. The summer generates the function

$$h(\Xi) = a_0 + \sum_{i=1}^{p} a_i \Xi_i.$$

The threshold element generates a +1 signal for $h(\Xi) > 0$ and a - 1 signal for $h(\Xi) < 0$ the signal is signum $\{h(\Xi)\}$. Signum $\{0\}$ may be arbitrarily defined as being + 1 rather than - 1.

planes (which are planes in hyperspace); the location of such surfaces is determined by the values of fairly few constants. The equation for a hyperplane is $h(\Xi)=0$ where

(12)
$$h(\Xi) = a_0 + \sum_{i=1}^{p} a_i \Xi_i$$

The $(p+1)$ constants $a_0, a_1, \ldots, a_p$ represent p pieces of informa tion. Any point on the positive side of the hyperplane has a pos itive value of $h(\Xi)$; any point on the negative side has a negative value of $h(\Xi)$. The location of a point with respect to a hyper- plane may therefore be determined from the sign of $h(\Xi)$; e.g., the sign of a_0 indicates on which side of the plane the origin is. The function signum $\{h(\Xi)\}$, or $\text{sgn}\{h(\Xi)\}$, may be generated by a threshold logic unit, TLU, as shown in Fig. 5. The TLU has been used alone as categorizer, it has also been used as the elemental building block of more elaborated categorizers, e.g., layered machines (Nilsson 1965, Chapter 6; Hoffman and Moe 1969) which are networks of interconnected TLUs and which include committee machines. A committee machine is a PR con sisting of an odd number of TLUs, each of which has one "vote", the PR makes decisions according to the majority of the TLU votes. Simple separation surfaces may be obtained by cascading TLU units as indicated by an example in Figure 6 (Vadzow 1968; Drucker 1969).

7.2.2 The Need for Separation Surfaces of Simple Shape.

The easiest way to illustrate the need is by way of a numerical example. Consider a problem with 2 classes, Class A and Class B, where the designer uses 50 attributes and where all 50 density functions look like Fig. 3. Clearly the 50-

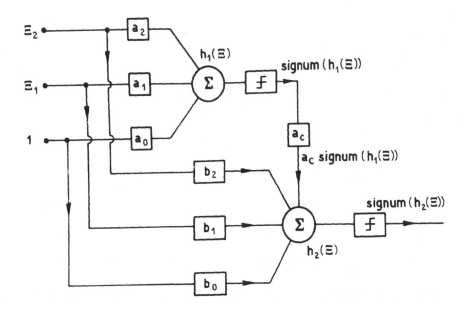

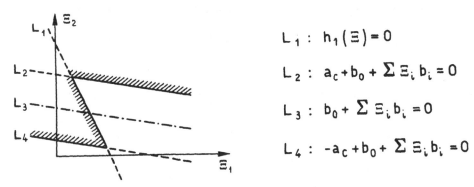

$$L_1 : h_1(\Xi) = 0$$

$$L_2 : a_c + b_0 + \Sigma \, \Xi_i \, b_i = 0$$

$$L_3 : b_0 + \Sigma \, \Xi_i \, b_i = 0$$

$$L_4 : -a_c + b_0 + \Sigma \, \Xi_i \, b_i = 0$$

FIGURE 6. The figure presents an example of how a nonlinear separation surface may be generated by scacading threshold logic units. Two TLUs and two attributes Ξ_1 and Ξ_2 are used. The heavy zig-zag line indicates the nonlinear separation surface in the (Ξ_1, Ξ_2)-plane. Above the zig-zag line $h_2(\Xi)$ is positive, below the line $h_2(\Xi)$ is negative. The equations for the four lines, L_1, L_2, L_3, and L_4, are listed above.

variate density function for Class A members may be anywhere

from trimodal to 3^{50}-modal and the density function for Class

B members may be anywhere from bimodal to 2^{50}-modal. The

separation surface obtained by 50 sets of 4 thresholds like those

in Fig. 3 will probably be suboptimal. The optimum separation

surface will in all likelihood be highly convoluted so the surface

will be difficult to locate and to store. In contradistinction to

this situation the designer should have little trouble finding an

optimum (or near-optimum) separation surface of simple shape

if he has 50 attributes like the one illustrated in Fig. 2.

The need for separation surfaces of simple

shape give rise to the question: what can the designer do to make

such surfaces optimum (or near-optimum) and in particular

what can he do to avoid that highly convoluted surfaces become

optimum? It is in general so that the surface which is needed

to separate disjointed classes (and to a lesser degree classes

with multimodal distributions) from each other is of a more

complicated shape than the surface needed to separate classes

with unimodal distributions. Therefore, the designer should

consider dividing any class with a multimodal distribution into

subclasses with unimodal distributions. There are two steps

which the designer can take to achieve such a decomposition.

(i) The designer should see if the members of the class in ques

tion, C_i , do actually come from distinct subclasses, $C_{i,1}$, $C_{i,2}$,

which are generically different. If so, the designer should con sider working with the subclasses rather than with C_i. E.g., C_i could be the class of typewritten capital A letters, $C_{i,1}$ could be Pica capital A letters, $C_{i,2}$ could be Elite capital A letters, etc. Membership in C_i will after the division of the class be decided in the last stage of the categorizer in the following manner. If the unlabelled pattern is identified as being a member of one (or more) of the C_i-subclasses and of none of the other $(N_c - 1)$ classes, the pattern is classified as being a member of C_i.

(ii) A second step the designer can take is to find the "Clusters", meaning locations in p-dimensional space with high densities of C_i-patterns. This can be achieved with suitable cluster seeking procedures (Sebestyen 1962, Section 4.5; Firschein and Fischler 1963; Mattson and Dammann 1965; Dammann 1966; Nagy 1968, Part IV; Gitman and Levine 1970; Haralick and Kelly 1969). Patterns that belong to cluster n° 1, 2 etc., are considered to belong to the subclasses $C_{i,1}$, $C_{i,2}$, etc. The designer can also determine the clusters by one of the methods for non-supervised classification of patterns (Cooper and Cooper 1964; Spragins 1966; Patrick and Hancock 1966; Fralick 1967; Tsypkin 1968). If the unlabelled pattern is identified as being a member of one (or more) of the C_i-subclasses and of none of the other $(N_c - 1)$ classes the pattern is classified as being a member of C_i.

It is necessary in practice to use separation

surfaces of simple shape. Fortunately, there are a number of important cases where the optimum form of the separation sur face actually is simple; examples of such cases will be present ed in Article 7.2.3. Sometimes the designer wants to separate two classes of pattern points having distributions of unknown forms. In such cases the designer may select a separation sur face of suitable but simple shape and locate it in an optimum fashion in pattern space; this procedure will be discussed in Article 7.2.4.

7.2.3. Parametric Training Methods.

Consider the case where the functional forms of the N_c multivariate densities are known. E. g. the members of C_i could have a multivariate Gaussian density. When the func tional forms of the N_c densities are known, the functional form of the optimum separation surface can be computed so as to op timize a specified IP. By parametric training methods are under stood methods that use the design data to estimate the values of the parameters for the computed optimum separation surface. The estimates are often said to be obtained from the design data by a "training procedure". These estimates are then later used as optimum values when the categorizer is realized in hardware or software. Two important cases will now be presented.

Separation Surfaces of Optimum Form. Let the

continually been adjusted so as to correct the erroneous classi
fications of the representative patterns. It gives a substantial
saving in computer time if the starting position of the hyperplane
is selected with care; if the members of both classes are uni-
modally distributed with centroids M_A and M_B, a reasonable
starting position would be one where the hyperplane bisects the
connecting M_A with M_B .

The ease with which M_{AD} points are separated
from M_{BD} points by a hyperplane obviously depends on the di
mensionality; this matter will now be discussed. The M_{AD} plus
M_{BD} pattern points in p-dimensional space are usually in "gen
eral position". A set of points is said to be in general position
when no subset of (p+1) points lies on a (p-1) dimensional hyper
plane. Let it be assumed that all the points are in general posi
tion and arbitrarily located in the p-dimensional space. Then
the points can be separated correctly (i) with 100% probability
when $M_{AD} + M_{BD} \leqslant p+1$, (ii) with more than 50% probability when
$M_{AD} + M_{BD} < 2p$, (iii) with 50% probability when $M_{AD} + M_{BD} = 2p$, and
(iv) with less than 50% probability when $M_{AD} + M_{BD} > 2p$. The change
in probability as a function of $(M_{AD} + M_{BD})/p$ is very pronounced at
the 50% level in particular for larger values of p (Nilsson 1965,
Figure 2.12). It consequently becomes natural to define the
capacity of a hyperplane as 2p or twice the number of degrees
of freedom, (Cover 1965). If the hyperplane can separate more
than 2p patterns correctly, and p is moderately large (say more

for the separation surface such as the linear, quadratic, or piecewise linear form (Nilsson 1965, Chapter 7); the designer's choice of form will depend on his expectations with regard to the modalities of the N_c densities. The functions have unspeci fied coefficients (often called weights), the values of which are adjusted by the designer in such a manner that the separation surface performs as well as possible on the design data. The coefficient adjustment is commonly referred to as training. If the members of two classes can be separated by a hyperplane they are said to be linearly separable. Non-parametric train- ing procedures for the "linearly separable" and "not linearly separable" case will be discussed followed by two notes on lin ear separation.

Linearly Separable Patterns. It is desired to find a hyperplane that separates M_{AD} representative pattern points belonging to Class A from M_{BD} pattern points belong- ing to Class B. It is known a priori that the patterns are lin- early separable. Besides the linear programming approach (Smith 1968) several training procedures are available by which a suitable hyperplane always can be located after a finite number of iterations (Rosenblatt 1960; Nilsson 1965, Sections 4.3 and 5.1; Ho and Kashyap 1965; Glucksman 1966); numerical exam- ples of such procedures are available in the literature (Nilsson 1965, Table 4.1; Nagy 1968, Table II). The procedures are call ed error-correcting because the location of the hyperplane is

p attributes have a multivariate, Gaussian or Pearson Type II,
or Pearson Type VII, density for both members of C_1 and mem
bers of C_2 ; in this case the optimum separation surface is a
quadric (Cooper 1964). The location of the quadric may be deter
mined by the two mean vectors and the two covariance or scal
ing matrices, all of which can be estimated from the design data.
Under special conditions the quadric reduces to two hyperplanes
or only one hyperplane. E.g., if two multivariate, Gaussian
distributions have the same covariance matrix the optimum sepa
ration surface is a single hyperplane (Cooper 1964).

Separation Surfaces of Suboptimal Form. In
case the optimum separation surface is too convoluted the designer may compute the optimum location of a separation surface
the designer accepts a loss in maximized IP-value to gain a
simpler categorizer implementation and a simpler (and maybe
faster) categorization procedure. An example of a sub-optimal
separation surface is a hyperplane used (instead of a quadric)
to separate two Gaussian, multivariate distributions with uneq-
ual covariance matrices (Anderson and Bahadur 1962).

7.2.4 Non-Parametric Training Methods.

Non-parametric training methods are applied
when no assumptions can be made about the N_c multivariate dis
tributions. In this situation some functional form is assumed

than seven), it indicates that the members of both classes "clus

ter" well. The existence of clusters again indicates that the p

attributes have been well chosen.

Patterns Which Are Not Linearly Separable.

It is desired to find a hyperplane that separates representative

pattern points from two classes, Classes A and B, while optim

izing a specified IP. No assumptions are made about the repre

sentative pattern points being linearly separable. This problem

has been studied and answers have been obtained which are ap-

plicable when Bayes' criterion can be used to measure the per

formance of the hyperplane (Highleyman 1962; Akers 1965; Wolff

1966; Chen 1969) or other IPs are used (Peterson and Mattson

1966); the answers take the form of computer programs, as an

example the Akers program will be described in Appendix 2.

Committee machines have also been used successfully to solve

problems of this kind (Nilsson 1965, Sections 6.2, 6.3 and 6.4).

The special case where the p attributes are all two-valued func

tions has received considerable interest and has been surveyed

by Winder (Winder 1969).

First Note. Consider a problem where the weight and the height

of the patterns have found to be two effective attributes, Ξ_1

and Ξ_2. Clearly the effectiveness of the attributes is indepen

dent on whether the height is measured in inches or miles and

whether the weight is measured in pounds or stones. In either

case the optimally located hyperplanes will give precisely the

the same performance although the attributes have been weighted differently; a_1 and a_2 in Equation 12 will be different. There is consequently in general no reason to believe that the attributes which are weighted heavily (those with high a-values) are more effective than the attributes with low weight values.

Second Note. Assume that the designer has found the optimum location for a hyperplane which separates Class A points from Class B points. The designer may now want to perform a linear transformation of pattern space to improve the clustering of Class A points and the clustering of Class B points. During the transformation the hyperplane is changed into a surface of the same functional form namely a new hyperplane. The points which were on opposite sides of the plane before the transformation remain on opposite sides after the transformation. The performance of the (optimum) plane is precisely the same before and after the transformation; the transformation may result in better clustering but it does not result in fewer misclassified pattern points. A transformation may improve the PRD performance in situations where the "post-transformation optimum separation surface of realizable form" corresponds to a "pre-transformation separation surface" of a convoluted form which the designer cannot (or does not want to) realize; transformations which lead to better PRD performance will often include folding in some form.

Some Other Methods. For the sake of complete

ness it should be mentioned that some simpler classification schemes (Sebestyen 1962, Section 5.1) such as "Correlation with Stored References" and "Proximity to Nearest Regression Plane" or "Nearest-Class-Mean Classification" also lead to separation surfaces consisting of hyperplanes. The different methods have been illustrated by numerical examples in Nagy 1968, Table 1 and Figure 2.

If each attribute can take only a few distinct values, it is possible to use a decision tree (e.g., Unger 1959 and Glucksman 1965) or Boolean functions (Ishida and Steward 1965; Appendix 1) in the categorizer but such methods have been reported only rarely in the literature. The reason is probably that the two methods are not well suited for separation of popu lations with overlapping distributions and populations that are described by representative members only.

Hardware Implementation. Several machines, special purpose computers, have been built, by which the adjusting of the separation surface coefficients may take place in hardware; e.g., the Perceptron (Rosenblatt 1960; Block 1962; Rosenblatt 1962), Adaline (Widrow et al. 1963), Minos (Brain et al. 1963), and the Learning Matrix (Steinbuch and Piske 1963; Kazmierczak and Steinbuch 1963; Steinbuch and Widrow 1966). The computing ability of Perceptrons has recently been studied with thought provoking results (Minsky and Papert 1969).

7.3 Implicit Partitioning of the Pattern Space.

This subsection is divided into two parts. In the first part, Article 7.3.1, nearest-neighbour-pattern classi fication is discussed. In the rest of the subsection different forms of discriminant functions are described. In Article 7.3.2 the equivalence of discriminant functions and separation sur- faces is explained. Next two methods for non-parametric design of discriminant functions are presented. Finally, in Article 7.3.6, parametric training of discriminant functions is discuss ed.

7.3.1 Nearest-Neighbour-Pattern Classification.

The nearest neighbour decision rule (Cover and Hart 1967; Cover 1968; Hellman 1970) assigns to an unlabelled pattern point the classmembership of the nearest of a set of previously classified pattern points. The rule is independent of the N_C underlying multivariate distributions, it is a non-para- metric categorization rule. Let the probability of misclassifying a pattern with this rule be called R_{NN} and with Bayes' proce- dure be called R^* (Bayes' procedure requires complete know- ledge of the underlying statistics). In the case where a large number of design data is available it may be shown (Cover and Hart 1967) that:

$$R^* \leqslant R_{NN} \leqslant R^*\left(2 - N_c R^* / (N_c - 1)\right) \tag{13}$$

In other words: when a large number of representative pattern points from the N_c classes can be stored and the one closest to the unlabelled point can be identified, then the probability of error with this procedure is less than $2R^*$. The method is intuitively appealing but it suffers from the shortcoming that it is difficult both (i) to store a large number of points and to compute the many distances among which the smallest is desired or (ii) to find the "minimal consistent subset" of correctly classified pattern points (Hart 1968) which will perform as well as the original large set of points.

7.3.2 Discriminant Functions and Separation Surfaces.

Any separation surface can be defined implicitly by a set of N_c functions, $g_i(\Xi) = g_i(\Xi_1, ..., \Xi_p)$, $i = 1, ..., N_c$. These functions, which are scalar and single-valued, are called discriminant functions. They have the property that $g_m(\Xi) >$ $> g_n(\Xi)$ for $m, n = 1, ..., N_c$, $m \neq n$, whenever a pattern point Ξ is located in a C_m-region; a C_m-region is a region in the p-dimensional decision space where it is decided to classify all points as being representative of members of class C_m. The N_c discriminant functions may be generated in many ways; when Bayes' procedure is used, one possible approach is for each micro-region to use minus one times the $N_c(K_i)_k$ -values from Equation 3, Subsection 6.1 as function values. Each discriminant may be illustrated by a surface in a (p+1)-dimensional space obtained by adding a g-axis to the p-dimensional decision space. The implicitly defined separation surface coincides with the projec tion on the p-dimensional decision space of the loci of points in (p+1)-dimensional space where the two highest values of the discriminant functions have the same value (Nilsson 1965, Fig. 1.3; Wolverton and Wagner 1969; Meisel 1969). When a set of N_c discriminant functions is known, it is consequently possible to locate the separation surface. If the separation surface is known, it is possible to determine infinitely many sets of N_c discriminant functions by following the two guidelines. (1) For any pattern point Ξ inside a C_m -region the N_c discriminant functions can take any desired values as long as $g_m(\Xi) > g_n(\Xi)$ for $m, n = 1, \ldots, N_c$, $m \neq n$. (2) For any point Ξ on the separation surface in an area where the surface separates a C_m -region from a C_n -region the N_c discriminant functions can take any desired values as long as $g_m(\Xi) = g_n(\Xi) > g_k(\Xi), k, m, n = 1, \ldots, N_c$, $n \neq m \neq k \neq n$. Consequently, it is seen that just as a set of N_c dis-criminant functions determines the separation surface, the sep-aration surface determines an infinite number of possible sets of N_c discriminant functions.

7.3.3 Categorization Using N$_c$ Discriminants.

The categorizer may be designed so that it uses N$_c$ discriminant functions rather than the expression for the separation surface. If so, the categorizer consists of (1) N$_c$ blocks where the values of the N$_c$ discriminant functions, $g_i(\Xi)$, are computed for the unlabelled pattern Ξ (the N$_c$ computations are often performed in parallel) followed by (2) a block, the maximum selector, where it is decided which discriminant function has the highest value. The index number of that function is then presented as the result of the categorization. When each of the $g_i(\Xi)$-functions is a linear function of $\Xi_1, \Xi_2, \ldots, \Xi_p$, the machine is said to be a linear machine (Nilsson 1965, Section 2.3).

The form of one of the N$_c$ functions may be chosen arbitrarily. E.g., if $\left(g_i(\Xi)-g_1(\Xi)\right)$ is ·used instead of $g_i(\Xi)$, $i=1,\ldots,N_c$, it is seen that $g_1(\Xi)=0$ for all Ξ -values. A pattern, Ξ , is then classified as being a member of Class 1 if $\left(g_i(\Xi)-g_1(\Xi)\right)<0$, $i=2,3, \ldots ,N_c$. With this procedure one of the N$_c$ discriminant function generators can be saved; for N$_c$ = 2 one discriminant function will suffice, this situation is encountered quite frequently in the literature.

A number of possible discriminant functions have been discussed in the literature (Nilsson 1965, Chapter 2). The most popular function is the linear discriminant function

g_{i_L} (Peterson and Mattson 1966).

(14) $g_{i_L}(\Xi) = c_{i0} + c_{i1} \cdot \Xi_1 + \ldots + c_{ip} \cdot \Xi_p$

This function is simple to implement in hardware with a summing circuit. Two logical extensions "the piecewise linear discriminant function" (Nilsson 1965, Chapter 6; Duda and Fossum 1966) and "the polynomial discrimination function" (Specht 1967b) have also received attention. The Φ-machine is a powerful method for generating linear discriminants, it will be described next.

7.3.4 The Φ -Machine.

Let $\Phi_j = \Phi_j(\Xi_1, \ldots, \Xi_p)$, $j=1, \ldots, d$, be a real-valued measurement function. Consider a machine that first computes the values of the d Φ -functions for a specific pattern point, secondly selects the largest of the N_C discriminants Φ_i, $i = 1, \ldots, N_C$, defined by Equation 15, and finally announces the index number of the discriminant as the classification of the pattern. The $(d+1) \cdot N_C$ quantities in Equation 15, $b_{i,j}$, $j=0,1,\ldots,d$, are coefficients which can be adjusted during the training.

(15) $\Phi_i = b_{i,0} + \sum_{j=1}^{d} b_{i,j} \cdot \Phi_j$, $i = 1, \ldots, N_C$

Such a machine which uses Φ_i functions as discriminants is called Φ -machine (Nilsson 1965, Section 2.11). Notice the similarity in the form of the expressions for a linear discrim inant, $g_{i_L}(\Xi)$, Equation 14, and Φ_i , Equation 15; $g_{i_L}(\Xi)$ and Φ_i are both linear functions of a set of adjustable coefficients. The Φ -machine has consequently the advantages of a linear machine meaning that it is simple to train non-parametrically; this property was mentioned in Article 7.2.4. But thanks to the computation of the Φ -functions, the Φ -machine is more power ful than a linear machine, which just selects the maximum of $g_{i_L}(\Xi), i = 1,..., N_C.$ E.g.if $\Phi_j = (\Xi_j - c_j)^2$ where c_j is a non-adjust able constant, $j = 1,..., p$, Equation 1.15 describes a quadratic surface; a quadratic surface clearly is a more powerful dis- criminant (Nilsson 1965, Section 2.12).

To keep the presentation in this report concept ually simple, assume that all $d\,\Phi_j$ -functions are computed in the last stage of the receptor. If so, d and Φ_j become equal to what has been (and will be) understood by p and Ξ_i , and Φ_i becomes a linear discriminant like $g_{i_L}(\Xi)$. Pattern classes which can be separated by a Φ -machine will in the following be called linearly separable.

7.3.5   The Nonlinear Generalized Discriminant.

A different way of applying the concept of a discriminant function is to use only one function, the "Nonlinear Generalized Discriminant" (Sebestyen 1962, pp. 131-134), in- stead of using N_C discriminant functions. The nonlinear generalized discriminant is a function, $U = U(\Xi_1,...,\Xi_p)$, which could

be plotted in a (p+1)-dimensional space obtained by adding a

U-axis to the decision space. The discriminant has the property

that its value over all C_i-regions in decision space fall in one

interval U_i on the U-axis, and that the N_C intervals on the U-

axis, U_1, U_2, etc. do not overlap. Based on the U-value for an

unlabelled pattern (or rather the U-interval in which it falls)

it may consequently be decided to which of the N_C classes the

pattern belongs. "In practice, however, the number of coef-

ficients that must be determined in finding a suitable approxima

tion of the nonlinear function becomes very large as the number

of classes, dimensions of the space, and degree of the approx

imating function are raised. The matrix that must be inverted

to find the optimum choice of unknown coefficients becomes

prohibitively large. It is for this reason that less sophisticated

methods must often be employed to solve practical problems",

(Sebestyen 1962, p. 133).

7.3.6 Parametric Training of Discriminants.

Sometimes the designer makes decisions us

ing the likelihood method, Subsection 6.4, or equivalent proce

dures. In such cases the categorizer may be implemented by

(i) a memory containing the N_C multivariate density functions

and (ii) a comparator which selects the class for which the mem bership is most likely given the Ξ -value of the unlabelled pattern. The N_C multivariate density functions are discriminants which must be determined by parametric training; they can be estimated by any of the methods described in Subsection 7.1.

A particular case with modest storage requirements deserves special mention. It is the case where the p attributes are all two-valued. The two values are called 1 and 0. Assume that attribute n° k, Ξ_k , takes the value 1 with probability $p_{k,i}$ for members of C_i , $i = 1,2,...,N_C$. Let it be assumed that the attributes are statistically independent. Under the independence assumption the multivariate density for members of C_i is $P_i(\Xi)$.

$$P_i(\Xi) = P_{pr}(i) \cdot \prod_{k=1}^{p} \left(p_{k,i}^{\Xi_k} \cdot (1-p_{k,i})^{1-\Xi_k} \right) \qquad (16)$$

$$\log\left(P_i(\Xi)\right) = \log\left(P_{pr}(i)\right) + \sum_{k=1}^{p} \log\left(1-p_{k,i}\right) + \Xi_k \cdot \log\, p_{k,i} / \left(1-p_{k,i}\right))),$$
$$(17)$$

$\log\left(P_i(\Xi)\right)$ increases monotonically with $P_i(\Xi)$; the N_C functions, $\log\left(P(\Xi_i)\right)$, $i = 1, ...,N_C$ can consequently be used as linear discriminants when the designer has knowledge of or can estimate the probabilities in Equation 17. Equation 17 shows that $\log\left(P_i(\Xi)\right)$ is a linear function of $\Xi_1, ..., \Xi_p$; such a function is simple to implement with a summing circuit.

7.4 Categorization of Members from More than Two Classes.

When the partitioning of pattern space is <u>implicit</u> as described in Subsection 7.3, the categorizer design is in principle the same for the cases $N_C = 2$ and $N_C > 2$. The only difference between the two cases is that a larger categorizer may be needed when N_C increases; e.g., the stored representative points represent more classes if "nearest neighbor classification" (Article 7.3.1) is used, and the number of a posteriori probabilities (one per class) that must be comput ed and compared in a conditional probability computer (Selfridge 1959) will also increase. When the partitioning of the pattern space is <u>explicit</u> and $N_C = 2$, it is convenient to locate members of one class of patterns on the "inside" of the separation surface and the members of the other class on the "outside". With $N_C > 2$ this simplicity is lost; the problem may, however, be solved by one of the following four methods for separation of N_C classes by use of hyperplanes. All four methods are based on the reduction of the N_C class problem to a set of two-class problems.

<u>Method n°1.</u> The N_C classes are separated pairwise (Highleyman 1962), this technique requires $N_C(N_C-1)/2$ or fewer hyperplanes. A pattern is considered a member of Class C_i when the corresponding point falls on the C_i-side of each of the (N_C-1) hyperplanes that separate members of C_i from members of the

other $(N_C - 1)$ classes.

Method n° 2. As a first step it is only attempted to separate members of two of the N_C classes, say C_a and C_b. The resulting hyperplane will in all likelihood divide the representative members of one or more of the remaining $(N_C - 2)$ unevenly, say 90% vs. 10%. Let it be assumed that almost all of the members of C_c fell on the positive side of the hyperplane just like the members of C_a and almost all of the members of C_d fell on the negative side of the hyperplane where the members of class C_b are located. Furthermore, let it be assumed that the members of each of the remaining $(N_C - 4)$ classes were divided evenly by the hyperplane, 50% vs. 50%. In a next step it will be attempted to separate members of C_a and C_c from members of C_b and C_d. Assume that the attempt succeeds and that the members of the remaining $(N_C - 4)$ classes still tend to be divided evenly by the plane. If so, the result may be illustrated by a column in a truth table with two 1-digits for C_a and C_c, two 0-digits for C_b and C_d, and $(N_C - 4)$ don't-care-statements; such a column marked Hp1 is shown in table 1. By continuing this procedure a set of S_H hyperplanes may be found that determines a truth table with $(S_H + 1)$ columns and N_C rows. Any two rows should differ with respect to the 1- or 0-entries in at least one column. A decision regarding class membership of an unlabelled pattern may now be reached in the following manner. First it is determined on which side of each of the S_H hyperplanes

TABLE   1

Class	Hp 1	Hp 2	Hp 3	Hp 4
a	1	1	0	d.c.
b	0	d.c.	1	0
c	1	0	d.c.	1
d	0	d.c.	0	1
e	d.c.	0	0	0
f	d.c.	1	1	1

Separation of the members of six classes by 4 hyperplanes

Method no. 2, Subsection 7.4 is used. Hp 1 means hyperplane no. 1, etc. ; d.c. illustrates the don't-care condition. As an example of the classification procedure consider a pattern that is located on the positive side of all four hyperplanes. The pattern is coded '111' and is classified as being a member of Class f. A pattern with code '1010' would be rejected as not belonging to any of the six classes in question.

the corresponding point is located. Location on the positive side is indicated by a logic 1, location on the plane or on the negative side is indicated by a logic 0. Secondly, by a suitable logic, the location of the point as expressed by a string of S_H binits is compared to the N_C rows and it is decided to which, if any, of the N_C classes the pattern belongs. A numerical example with $N_C = 6$ is presented in Table 1.

This second method requires fewer hyper- planes than the first method. The value of S_H is bounded by the following two expressions in N_c .

$$N_c \left(N_c - 1\right)/2 \geqslant S_H \geqslant \log_2 \left(N_c\right) \tag{18}$$

It should be realized that the second method requires some plan

ning of the trial-and-error procedure, whereas no such effort

is needed with methods n° 1, 3 or 4.

Method n° 3. The N_c classes are separated by N_c hyperplanes.

With this technique hyperplane n° i should be so located that it

on one side has the members of Class C_i and on the other side

the members of the other $\left(N_c - 1\right)$ classes. The method may be

considered when the representative members of each class

form a tight cluster and the $N_c \left(N_c - 1\right)/2$ distances between clus-

ters are all large compared to the largest "cluster-radius".

In that case each of the N_c hyperplanes should be located so that

it separates the N_c "points of finite size" in the manner just

described. If the designer can implement hyperspheres or ortho

topes (the p-dimensional equivalent of a 3-dimensional box),

he may consider the possibility of locating the members of each

class inside a particular closed surface. This form of catego

rizer has the particular advantage that the addition of a new

class, Class n° $N_c + 1$, only may require the addition of a new

closed surface rather than a complete redesign of the catego-

rizer (Bonner 1966).

Method n° 4. Reformulations as a two-class problem. It is de
sired to find a set of N_C separation surfaces which can separate
the pattern points representative of $C_i, i = 1, \ldots, N_C$, from the
pattern points representative of the remaining $(N_C - 1)$ classes.
It is known a priori that the representative pattern points from
any two classes are linearly separable. It is possible to reduce
this problem to a two class problem in a space with more than
p dimensions by a reformulation. With this technique a set of
N_C hyperplanes can always be located after a finite number of
iterations (Nilsson 1965, Sections 4.5 and 5.5). In the case
where the N_C classes are not separable related techniques may
be used (Chaplin and Levadi 1967; Wee and Fu 1968).

It has been shown in this subsection that the
problem of separating members of N_C classes can be reduced
to the problem of separating members of two classes. In the
remaining part of this report it will be assumed, unless other
wise stated, that $N_C = 2$, so that only two pattern classes are
of concern.

8. Hardware Implementations

Besides constraints on funding and on time available for job completion, the designer must keep other lim-iting factors in mind during the receptor and categorizer design, especially if the design should be realized in hardware. The designer will usually want the operations in the PR to have the following two properties.

(1) A reliable and economical implementation of the operations should be possible. This means that certain esoteric functions may have to be ruled out although they could have been used to measure the values of interesting attributes. Functions that can be realized with digital circuitry will usually be found to give promises of high reliability because such circuitry is well suited for micro-miniaturization. There are certain applications where volume and weight are of more than usual concern, e. g., space- and medical electronics. In such areas schemes which at first may seem a bit far fetched, such as the para-propagation concept (Glucksman 1965), may become attractive on account of hardware that is made up of micro-circuit mod-ules (Idzik et al. 1964). The FOBW-technique that is described in Appendix 4 does also result in a potentially reliable hardware implementation as will be seen. Often a PR will be located in a hostile environment because the integration of a sensor with a PR promises a substantial reduction in data handling. In such

cases circuit design problems can be expected to influence the PR design. E.g., when the ambient temperature is high it becomes necessary to use design methods that will maximize the probability of not having the system fail due to drift failure or to catastrophic failure (Becker and Warr 1963).

(2) The results should be available quickly and without use of excessive memory. How critical these requirements are changes. In many cases, however, the designer may find it to his advantage to use schemes where (i) the attribute values may be computed in parallel rather than in series and where (ii) the unlabelled pattern can be processed little by little as it is presented to the receptor (Bonner 1966), rather than after the complete pattern has been stored in a buffer memory. In Appendix 4 it will be shown that the FOBW technique requires only little memory: a shift-register to recall the immediate past of a binary sequence with precision, and p counters of some kind (one per attribute) to generate the frequencies of occurrence of a certain selected binary words; the contents of the counters give a statistical description of the more distant past of the binary sequence. The p binary word frequencies are generated in parallel and are quickly available for processing in the categorizer.

Building PRs by simulating the nervous system is potentially a very powerful approach. It seems, however, that substantial advances in the field of integrated circuitry are required before the full potential of the method can be real ized. The requirement of a large-capacity semiconductor memory may not be too serious (Hodges 1968). The real stumbling -block seems to be the lack of a simple electronic equivalent of a neuron that can be massproduced in net-configurations. Substantial work has been done, nevertheless, in modelling nervous systems (Carne 1965; Deutsch 1967). Also the modelling of neurons has been an active field; an important problem here is to obtain recoverable neuristor propagation (Kunov 1967; Parmentier 1970), the neuristor being the electronic equivalent of the nerve axon. A review of the work in the two areas has been given in an article by Harmon and Lewis (Harmon and Lewis 1966) and recent progress has been reported in the Special Issue on "Studies of Neural Elements and Systems", The Proceedings of the IEEE, June 1968.

APPENDIX 1

A PRD for Recognition of Handprinted Arabic Numerals (Nielsen 1970)

1. Introduction.

The need for devices which can recognize hand printed symbols is encountered in a number of areas related to computer technology. Such devices can speed up the data preparation and input operation, facilitate the transmission of writ ten statements and signatures over telephone lines, expedite mail sorting (Chenchi et al. 1968), simplify certain language translation tasks, and facilitate problem solving using machine interaction. The design of such devices has been the subject of considerable research in the past decade; a paper by Munson (Munson 1968) is of particular interest in this regard as it dis cusses more than 40 papers, published before 1968, on ways of recognizing handprinted characters.

2. A Topic for a Masters Thesis.

It was felt at the Electronics Lab., Technical University of Denmark that the building of a device for automatic recognition of handprinted arabic numerals would constitute a suitable Masters Thesis Project. The project was undertaken and carried out by Mr. Knud Arne Nielsen (Nielsen 1970).

3. Recognition of Handprinted Digits.

This is a problem which has received consider able attention over the years so the state of the art is well known. We decided to solve the problem by (1) dividing the tablet, on which the writing is done, into zones which are carefully match ed to ten pattern classes, and (2) using a recognition logic, to be described in the following section, which exploits the dynamic information available. The approximate shape of the zones was determined by the knowledge we have accumulated in our minds of what the ten digits look like. The 20 representative sets of 10 digits were obtained from students and the precise zone-loca tions were determined so as to minimize the mean classification error. It was not deemed necessary to evaluate the design with a set of "test data" so all the data used as "design data". Figure 1A shows the zone-locations. The procedure was cut-and-try, and we disregarded certain problem areas. For example, we just asked the students to write in a "natural manner" and dis regarded the interesting human factors associated with careful handprinting; also, we made the dubious assumption that each digit has a prior probability of 10%, when we minimized the mean classification error (Hamming 1970).

4. The Recognition Logic.

The digits are print<u>ed</u> on a copper plate with a coppertip-ped pen which is electrically grou<u>nd</u>ed. The copper plate is divided into zones each of which is electrically insulated from the adjacent zones. Each zone is indicated by a lower case letter, a, b, c, etc.; zones with the same letters are connected externally. When the zone "a" is touched by the pen a flip-flop called

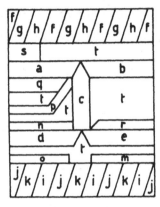

Fig. 1 A. The copper plate. The dark lines indicate insulation between zones. Zones with the same lower case letters are connected externally. The actual size of the plate is 45 mm by 55 mm.

A, changes state from "0" to "1". Likewise, flip-flops named B, C, D, etc. do change their state from "0" to "1" whenever the pen happens to touch a zone indicated by the corresponding lower case letter. The only exception to this rule is the set of five zones marked "t"; these zones are used solely to ascer<u>tain</u> that writing is taking place. In the following the "1" states of the flip-flops will be referred to as A, B, C, etc., and the "0" states of the flip-flops will be referred to as $\bar{A}$, $\bar{B}$, $\bar{C}$, etc.

The actual recognition of the handprinted digits

is achieved partly through use of "dynamic information", meaning the observed values, "0" or "1", of certain "dynamic attributes" (Section 5), and partly through use of "static informa
tion" meaning the observed values of certain "static attributes"
(Section 6). The recognition logic is based solely on Boolean
functions; incidentally, Boolean logic is rarely used in the field
of Pattern Recognition.

5. The Dynamic Attributes.

          The most important of the three dynamic at
tributes is called PL, for "pen lifted". Whenever during the
printing of a digit the pen is lifted from the plate for a period
less than some specified fraction of a second, called T seconds,
PL changes value from "0" to "1". If the pen remains lifted for
more than T seconds it is assumed that the complete digit has
been printed and a decision is made and displayed with regard
to what digit was printed. Whenever the digit 2 has been print
ed with care PL remains "0". Whenever the digit 4 has been
been printed with care PL has changed to "1". Whenever the
digit 8 has been printed PL may be "0" or "1" depending on the
manner of printing. The values of the two remaining attributes,
the Boolean functions, F1 and F3 depend on PL. F3 and F1
will now be defined.
F3 changes value from "0" to "1" if the pen touches any of the

zones marked f, g, or h after PL changed from "0" to "1".
After PL has changed to "1" the flip-flops are disconnected
from the f, g and h zones so by now F, G, and H cannot change
state. This means that the attribute F1 defined below as Boolean
function of the same name becomes a dynamic attribute.

$$F1 = \overline{F \cdot G \cdot H} \qquad\qquad (1)$$

6. The Static Attributes.

The static attributes are the settings of the
flip-flops after the pen has been lifted from the plate for more
than T seconds. If at time J, K, and L all are "1" it indicates
the presence of a "horizontal" line at the bottom of the digit;
the attribute F2 defined as a Boolean function of the same name
then takes the value "0".

$$F2 = \overline{J \cdot K \cdot L} \qquad\qquad (2)$$

Similarly Equation (1) shows that F1 takes the value "0" in case
a "horizontal" line at the top of the digit was drawn before the
pen was lifted. To use F1 and F2 we must insist that the digit
is of a reasonable size and is reasonably well centered; this is
achieved by insisting that

$$R\,2\,(R1+F3)="1"$$

where the Boolean functions Rl and R2 are defined below.

(3) $$R1 = \overline{\overline{F} \cdot \overline{G} \cdot \overline{H}}$$

(4) $$R2 = \overline{\overline{J} \cdot \overline{K} \cdot \overline{L}}$$

7. The Decision Functions.

We are now in a position to list the decision functions; the handprinted digit is recognized in the following manner. Whenever the Boolean function Bl,

(5) $$B1 = R1 \cdot R2 \cdot F1 \left(\overline{F2} \cdot \overline{F3} \cdot PL + F2 \cdot \overline{PL} \right)$$

takes the value "1", "the digit one" is recognized. Here, as in the following cases, the corresponding identification light is turned on immediately after identification.
Whenever the function B4,

(6) $$B4 = R2 \cdot F1 \cdot F2 \cdot PL \left(R1 + F3 \right)$$

takes the value "1", the "digit four" is recognized.
Whenever the function B5,

(7) $$B5 = R1 \cdot R2 \cdot F1 \cdot \overline{F2} \cdot F3 \cdot PL$$

takes the value "1", "the digit 5" is recognized.

When the pen has been lifted from the copper plate for more than T seconds, each of the 16 attributes, F1 to S, listed in Table 1, will have taken on the value "0" or "1". If the set of 16 binary digits matches one of the 31 rows in Table 1, the handprinted digit is recognized as the corresponding digit listed in the leftmost column.

A hand printed digit is rejected and the "rejected" light is turned on in the following two cases. (1) When

$$R2\left(R1+F3\right)="0",$$

where R1 and R2 are defined by Equations (3) and (4), the size of the digit is too small. (2) If the values of the 16 attributes do not result in recognition, i.e. the values of B1, B4 and B5, defined by Equations (5), (6) and (7) have all been computed as "0" and furthermore, Table 1 has been consulted to no avail. It should be noticed that when a digit has been printed one and only one light will be turned on.

8. The Hardware Implementation.

The device was realized in hardware. Its performance is quite respectable; it recognized digits with about 90% accuracy. The device is illustrated in Figure 2A. The cost of the materials used was $150; if produced in quantity, the cost per device could be drastically reduced.

Fig. 2A
The device for recognition of handprinted digits.
The coppertipped pen is seen in the foreground.

Digit	F1	F2	F3	PL	A	B	C	D	E	M	N	O	P	Q	R	S
0	0	0	0	0	1	1	0	1	1	–	1	–	1	–	–	–
2	0	0	0	0	0	1	1	1	0	–	–	–	–	–	–	–
2	0	0	0	0	0	1	1	1	1	–	0	–	–	–	0	–
2	0	0	0	0	1	1	1	0	1	0	–	–	0	–	–	–
2	0	0	0	0	1	1	1	1	0	–	–	–	0	–	–	–
2	0	0	0	0	0	1	0	–	1	0	–	–	–	–	–	–
2	0	0	0	0	–	1	0	0	1	0	–	–	–	–	–	–
2	0	0	0	0	1	0	–	1	0	–	–	1	–	–	–	0
2	0	0	0	0	1	1	–	1	1	0	0	–	0	–	–	–
3	0	0	0	0	0	1	1	0	1	–	–	–	–	–	–	–
3	0	0	0	0	0	1	1	1	1	–	0	–	–	–	1	–
3	0	0	0	0	1	1	1	0	1	1	–	–	0	–	–	–
3	0	0	0	0	0	1	0	–	1	1	–	–	–	–	–	–
3	0	0	0	0	–	1	0	0	1	1	–	–	–	–	–	–
3	0	0	0	0	1	0	–	1	0	–	–	0	–	–	–	0
3	0	0	0	0	1	1	–	1	1	1	0	–	0	–	–	–
3	0	0	0	0	1	0	1	–	1	–	0	–	–	–	–	–
6	–	0	0	0	0	1	1	1	1	–	1	–	–	–	–	–
6	–	0	0	0	1	0	0	1	1	–	–	–	–	–	–	–
6	–	0	0	0	1	0	1	1	1	–	1	–	–	–	–	–
7	0	1	0	0	1	1	1	0	1	–	–	–	0	–	–	–
7	0	1	0	1	1	1	1	0	1	–	–	–	–	–	–	–
7	0	1	0	–	1	1	1	1	0	–	–	–	–	0	–	–
7	0	1	0	–	1	1	1	1	1	0	0	–	0	–	–	–
7	0	1	0	–	–	1	0	0	1	–	–	–	–	–	–	–
7	0	1	0	–	0	1	1	1	–	–	–	–	–	–	–	–
7	0	1	0	–	1	0	–	1	0	–	–	–	–	–	–	0
8	0	0	0	–	1	1	1	1	1	–	1	–	–	–	–	–
9	0	–	0	0	1	1	1	1	0	–	–	–	1	1	–	–
9	0	–	0	0	1	0	–	1	0	–	–	–	–	–	–	1
9	0	–	0	0	1	1	1	–	1	–	0	–	1	–	–	–

TABLE 1 A.   Binary logic. Whenever after the digit has been printed the 16 binary attributes, F1 to S, have the values listed in one of the 31 rows the digit is recognized as the digit listed in the leftmost column. A dash , – , indicates a don't care condition.

APPENDIX 2

A Categorization Procedure by Akers

The following procedure has been developed by Akers (Akers 1965). In the form of a computer program the procedure can serve as the means for categorization, Block 13 in the block diagram. With this procedure separation of the Class A and Class B points by a hyperplane is attempted using a pattern-space with a dimensionality that is as low, or almost as low as possible. An important by-product of the categorization is that attributes may be detected that have little or no incremental effectiveness; a discussion of effectiveness was presented in Subsection 5.3. The procedure is as follows. First separation is attempted in a one-dimensional space (using only one of the p-attributes). The M_{AD} plus M_{BD} points are read in one by one and located in the one-dimensional space. The one-dimensional hyperplane, a point, is continually adjusted to give 100% correct separation. Assume that a point, say P_{17}, cannot be classified correctly. The program will then examine the remaining (p-1) attributes to see if the introduction of a second attribute (meaning the addition of a dimension to pattern space) will make 100% correct separation possible for P_{17} plus the points which previously had been read in. This will usually be the case. If so, more points are read in, and the two dimensional hyperplane, a straight line, is continually adjusted to give 100% correct separation. Assume that a point, say P_{57}, cannot be classified correctly. The program will then examine the remaining (p-2) attributes to see if the introduction of a third attribute (meaning the addition of a dimension to pattern space) will make 100% correct separation possible for P_{57} plus the points which previously had been read in. In this manner, the program introduces dimension after dimension when needed. If the Class A and Class B points are linearly separable the program will find a separation plane that utilizes few (though not necessarily the minimum number of) attributes. If the Class A and Class B points are not linearly separable in p-dimensions, the program will switch to a second mode. Now all points are read in, and an attempt is made to locate the hyperplane in such a manner that the number of misclassified points is made as small as possible. In this second mode, all p attributes are used.

In most practical cases it becomes necessary to use the second mode in the Akers program. When the second mode is used it should be noticed which attributes are introduced last. Such attributes, on the average, may be assumed to be less effective than the attributes used earlier. Usually the attributes used last are discarded. In this manner some insurance is obtained against having to implement a categorizer with ineffective attributes.

APPENDIX 3

A PRD for Recognition of Spoken Words (Thamdrup 1969)

In this Appendix a spoken-word-recognizer for 16 Danish words is described; the device was designed and built by Mr. Jan E. Thamdrup for his Masters Thesis (Thamdrup 1969). The purpose of the description is to present readers with an illuminating example of a PRD.

1. The Performance of the PRD.

The PRD recognizes words from a vocabulary of 16 Danish words suited for oral communication with a computer: the ten digits (nul, et, to, tre, fire, fem, seks, syv, otte, ni) plus the words "start", "stop", "plus", "minus", "og" (meaning: "and" in English), "komma". The number of pattern classes, N_c, is thus 16. When the 16 words are equiprobable in the dictated messages the device is 94,7% correct with Mr. Thamdrup as speaker and about 2/3 correct with randomly selected male Danish speakers. The performance of the device may also be described as follows. If the device recognizes the 16 words 100% correctly then each correctly classified word represents 4 bits of information. Due to equivocation (Bremermann 1968) which is unavoidable in practice each (correctly or incorrectly) recognized word represents less than 4 bits of information. On the average a word spoken by Mr. Thamdrup represents after recognition 3, 73 bits; if the male speaker is selected at random a recognized word represents about 3 bits (Thamdrup 1969, Figure 63.1).

2. The Design of the PRD.

The microphone output is simultaneously sent through 4 different channels each beginning with a filter. The filters have 3dB band-widths which are 2 octaves and the following center frequencies in cps: 380, 1.300, 4.000 and 10.000 the skirts are in all cases 18 db/octave. Each filter-output is rectified and a weight average of the immediate past is formed. Each of the 4 averages is continually being compared to both a low and a high threshold, i.e. each filter-output is quantized into three levels. In the terms from Figure 1: the receptor output is the set of values of four, three-valued attributes, Ξ_1 Ξ_2, Ξ_3 and Ξ_4. As the speech signal changes with time the receptor output, Ξ, changes between the $3^4 = 81$ points in a four dimensional space. In the categorizer 17 such transitions are listed which are particular for certain segments of the 16 words. Based on the observed transitions and their sequence the PRD classified the spoken word by turning on the corresponding indicator light (it may happen that no light or several lights

are turned on). The device is built on 27 printed circuit boards
mounted on a rack as shown in Figure 3A, the cost of the com‾
ponent was $ 200.

Fig. 3A
Mr. Thamdrup and the author with the word-recognizer.

APPENDIX 4

The FOBW Method

1. Introduction.

In this Appendix one possible answer is given to a problem which may be stated in the following manner. How should a PRD be designed: (1) which can typically perform tasks such as recognizing sounds generated by engine malfunctions (Page 1967), or recognize normal and abnormal physiological signals (such data is for instance transmitted during space flights)? (2) which should use circuitry amenable to micro-miniaturization? (this is desired in a number of areas e. g. space- and medical electronical)? (3) so that a practical, heuristic method for searching for effective attributes can be used (finding such a method is incidentally one of the major problems in the pattern recognition field)?

2. Some Considerations.

Each of the three parts of the problem statement give some guidance towards the solution. The representative patterns are typically: (1) corrupted by noise, and (2) without a well-defined beginning or end; it is therefore deemed advisable to look for attributes which are of a statistical nature.

The microminiaturized circuit functions which are commercial

ly available are usually digital (rather than analog) functions.

During design of the PRD it is consequently inviting to sample

each of the representative patterns and code each sample by

a logic "1" or "0" (or short binary word). The set or repre-

sentative patterns typical for the pattern classes in question

may in this manner be converted to a set of (usually quite long)

strings of binary digits. The search for an effective set of at

tributes is based on an observation which will now be described

briefly. Assume that representative patterns belonging to Pat

tern-Classes P_1, P_2, etc. have been sampled and coded in binary

form. Assume furthermore that certain short binary words A_1,

A_2, etc. tend to have higher frequencies of occurrence $(f_{A_1},$

f_{A_2}, etc.) among the members of Class P_j than among the

members of any of the other classes. When the "Frequency of

Occurrence of Binary Words"-method (hereafter called FOBW

method) is used all pattern attributes are such frequencies of

occurrence of selected binary words. A_1 could be the six binit

word "110011", A_2 could be "100111", A_3 could be a logic "1"

followed 7 sampling intervals later by a logic "0" (this binary

word is written as "1-7-0"), A_4 could be "0-5-1", and A_5 could

be "1-12-1". If so A_1, A_2, etc. can be used to generate some

what longer binary words B_1, B_2, etc. which contain the shorter

words A_1, A_2, etc. E.g. B_1 could be the seven binit word

"1100111" which contains A_1 and A_2, B_2 could be a logic "1" fol lowed 7 sampling intervals later by a logic "0" which again is

followed 5 sampling intervals later by a logic "1" (notice how

this word "1-7-5-1", contains A_3, A_4 and A_5). It can be

shown under fairly general circumstances that on the average

the frequencies of occurrence of the longer words (f_{B_1}, f_{B_2},

etc.) are more effective attributes than f_{A_1}, f_{A_2}, etc. The

procedure is now to examine the effectiveness of f_{B_1}, f_{B_2},

etc.; the binary words which actually have frequencies of oc-

currence that separate the classes well are used to generate

somewhat longer binary words C_1, C_2, etc. It may again be

shown that the frequencies of occurrence (f_{C_1}, f_{C_2}, etc.) of

these longer words on the average are more effective attributes

than f_{B_1}, f_{B_2}, etc.   Again the effectiveness of f_{C_1}, f_{C_2},

etc. are examined etc. etc. The iterative procedure is illustrat

ed by the feedback loop consisting of Blocks 12, 14 and 16 in the

block diagram. The designer follows the procedure until the

problem is solved or seems unsolvable.

BIBLIOGRAPHY

[1] Abend K., Harley T. J. and Kanal L. N.: "Classification of Binary Random Patterns", IEEE Trans. on Inf. Theory, vol IT-11, n° 4, pp. 538-544, Oct. 1965.

[2] Abramson N. and Braverman D.: "Learning to Recognize Patterns in a Random Environment", IRE Trans. on Inf. Theory, vol. IT-8, n°5, pp. 58-63, Sept. 1962.

[3] Akers Jr. S. B.: "Techniques of Adaptive Decision Making", TIS R65LS-12 (Internal Publication, General Electric Company), 1965.

[4] Akers Jr. S. B. and Rutter B. H.: "The Use of Threshold Logic in Character Recognition", Proc. IEEE, vol. 52, n° 8, pp. 931-938, Aug. 1964.

[5] Amari S.: "A Theory of Adaptive Pattern Classifiers", IEEE Trans. on Electr. Comp., vol. EC-16, n° 3, pp. 299-307, June 1967.

[6] Anderson T. W.: "An introduction to Multivariate Statistical Analysis". New York, John Wiley & Sons Inc., 1958.

[7] Anderson T. W. and Bahadur R. R.: "Classification into Two Multivariate Normal Distributions with Different Covariance Matrices", Annals of Mathematical Statistics vol. 33, pp. 420-431 June 1962.

[8] Anderson W. W.: "Optimum Estimation of the Mean of Gaussian Processes", Proc. IEEE, vol. 53, n°10, pp. 1640-41, Oct. 1965 (Correspondence).

[9]   Arley N. and Buch K.R.: "Introduction to the Theory
        of Probability and Statistics" New York, Sci
        ence Editions (Paperback edition), 1966.

[10]   Bakis R., Herbst N.M. and Nagy G.: "An Experimental
        Study of Machine Recognition of Hand-Printed
        Numerals", IEEE Trans. on System Science
        Cybernetics, vol. SSC-4, n° 2, pp. 119-132,
        July 1968.

[11]   Becker P.W. and Warr R.E.: "Reliability vs. Compo
        nent Tolerances in Microelectronic Circuits",
        Proc. IEEE, vol. 51 n° 9 pp. 1202-1204 Sept.
        1963.

[12]   Becker P.W.: "Recognition of Patterns Using the Fre
        quencies of Occurrence of Binary Words",
        Copenhagen, Denmark, Polyteknisk Forlag,
        1968

[13]   Block H.D.: "The Perceptron: A Model for Brain Func
        tioning, I", Reviews of Modern Physics, vol.
        34, pp. 123-135, Jan. 1962.

[14]   Bongard M.: "Pattern Recognition", London, Macmillan
        1970.

[15]   Bonner R.E.: "On Some Clustering Techniques", IBM
        Journal of R and D, vol. n°1 pp. 22-32, Jan. 1964.

[16]   Bonner R.E.: "Pattern Recognition with Three Added
        Requirements".IEEE Trans. on Electr. Comp.
        vol. EC-15, n° 5 pp. 770-781, Oct. 1966.

[17]   Borch Karl: "Economic Objectives and Decision Prob-
        lems", IEEE Trans. on Systems Science and
        Cybernetics, vol. SSC-4 n°3 pp. 266-270 Sept.
        1968.

[18]   Brain A.E., Forsen G., Hall D. and Rosen C.: "A Large
        Self-Contained Learning Machine", WESCOW,
        Paper n° 6.1, Aug. 1963.

[19] Brain A. E., Hart P. E. and Munson J. H. : "Graphical-
      Data-Processing Reserach Study and Experi-
      mental Investigation" Tech. Rept. ECOM-01901
      -25, Stanford Research Inst. Menlo Park, Calif.
      Dec. 1966.

[20] Bremermann H. J.: "Pattern Recognition, Functionals
      and Entropy", IEEE Trans. on Bio-Medical
      Engineering, vol. BME-15, n° 3 pp. 201-207,
      July 1968.

[21] Brick D. B. : "On the Applicability of Wiener's Canonical
      Expansions", IEEE Trans. on Systems Science
      and Cybernetics, vol. SSC-4 n° 1, pp. 29-38,
      March. 1968.

[22] Brousil J. K. and Smith D. R. : "A Threshold Logic Net-
      work for Shape Invariance", IEEE Trans. on
      Electr. Comp. vol. EC-16, n°6, pp. 818-828,
      December 1967.

[23] Caceres Cesar A. and Dreifus Leonard S. : "Clinical
      Electrocardiography and Computers ", New
      York: Academic Press, 1970.

[24] Cadzow James A. : "Synthesis of Nonlinear Decision
      Boundaries by Cascaded Threshold Gates",
      IEEE Trans. on Comp. vol. 17, n° 12 pp. 1165
      -1172, Dec. 1968.

[25] Calvert T. W.: "Nonorthogonal Projections for Feature
      Extraction in Pattern recognition" IEEE Trans.
      on Comp. vol. C-19, n°5, pp. 447-452, May 1970.

[26] Cardillo G. P. and Fu K. S.: "Divergence and Linear
      Classifiers for Feature Selection", IEEE Trans.
      on Automatic Control, vol. AC-12, n° 6, pp.
      780-781, Dec. 1967 (Correspondence).

[27] Carne E. B. : "Artificial Intelligence Techniques", Wash
      ington D. C. Spartan Division (Books Inc.)1965.

[28]   Casey R.G.: "Moment Normalization of Handprinted
       Characters", IBM Journal of Research and
       Development, vol. 14, n°5 pp. 548-557, Sept.
       1970.

[29]   Casey R. and Nagy G.: "Recognition of Printed Chinese
       Characters", IEEE Trans. on Electr. Comp.,
       vol. EC-15, n° 1 pp. 91-101, Feb. 1966.

[30]   Chaplin W.G. and Levadi V.S.: "A Generalization of the
       Linear Threshold Decision Algorithm to Multi
       ple Classes" in Tou 1967, pp. 337-355.

[31]   Chen C.H.: "A Note on a Sequential Decision Approach
       to Pattern Recognition and Machine Learning
       Information and Control, vol. 9 n°6, pp. 549-
       562, Dec. 1966.

[32]   Chen C.H.: "A Theory of Bayesian Learning Systems",
       IEEE Trans. on Systems Science and Cyber-
       netics, vol. SSC-5, n°1, pp. 30-37, 1969.

[33]   Chien Y.T. and Fu K.S.: "Selection and Ordering of Fea
       ture Observations in a Pattern Recognition Sys
       tem" Information and Control, vol. 12 n° 5/6
       pp. 394-414, May-June 1968.

[34]   Chow C.K.: "An Optimum Character Recognition Sys-
       tem Using Decision Functions", IRE Trans. on
       Electr. Comp., vol. EC-6 n° 4, pp. 247-254,
       Dec. 1957.

[35]   Chow C.K.: "A Class of Nonlinear Recognition Proce-
       dures", IEEE Trans. on Systems Science and
       Cybernetics, vol. SSC-2, pp. 101-109, Dec.
       1966.

[36]   Chow C.K. and Liu C.N.: "Approximating Discrete Prob
       ability Distributions with Dependence Trees",
       IEEE Trans. on Inf. Theory, vol. IT-14, n°3
       pp. 462-467, May 1968.

[37] Chow C. K.: "On Optimum Recognition Error and Reject
      Trade off", IEEE Trans. of Inf. Theory, vol.
      n° 1 pp. 41-46, Jan. 1970.

[38] Chuang P. C.: "Recognition of Handprinted Numerals
      by Two-Stage Feature Extraction", IEEE
      Trans. on System Science and Cybernetics, vol.
      SSC-6 n°2 pp. 153-154, April 1970 (Correspon
      dence.)

[39] Cochran W. T., Cooley J. W., Fawin D. L., Helms H. D.,
      Kaenel R. A., Lang W. W., Maling Jr. G. C.,
      Nelson D. E., Rader C. M. and Welch P. D.:
      "What Is the Fast Fourier Transform?" Proc.
      IEEE vol. 55 n°10 pp. 1664-1674, Oct. 1967.

[40] Collins N. L. and Michie D. editors: "Machine Intelli-
      gence 1", London, U.K., Oliver and Boyd
      Ltd. 1967.

[41] Cooper P. W.: "Hyperplanes, Hyperspheres and Hyper
      quadrics as Decision Boundaries" in Tou and
      Wilcox 1964, pp. 111-138.

[42] Cooper D. B. and Cooper P. W.: "Non-supervised Adap-
      tive Signal Detection and Pattern Recognition",
      Information and Control, vol. 7 n° 3 pp. 416-
      444, Sept. 1964.

[43] Cover T. M.: "Geometrical and Statistical Properties
      of Systems of Linear Inequalities with Applica
      tions in Pattern Recognition", IEEE Trans.
      on Electr. Comp., vol. EC-14 n°6 pp. 326-334,
      June 1965.

[44] Cover T. M. and Hart P. E.: "Nearest-Neighbor-Pattern
      Classification", IEEE Trans. on Inf. Theory,
      vol. IT-13, n° 1, pp. 21-27, Jan. 1967.

[45] Cover T. M.: "Estimation by the Nearest Neighbor Rule",
      IEEE Trans. on Onfo. Theory, vol. IT-14, n°
      1, pp. 50-55, Jan. 1968.

[46] Crook M. N. and Kellogg D. S.: "Experimental Study of
      Human Factors for a Handwritten Numeral
      Reader", IBM Journal of Research and Devel
      opment, vol. 7, n° 1 pp. 76-78, January 1963.

[47] Dale E. and Michie D., editors: "Machine Intelligence
      2", London, Oliver and Boyd Ltd., 1968.

[48] Dammann J. E.: "An Experiment in Cluster Detection",
      IBM Journal of R and D, vol. 10 n° 1 pp. 80-
      88, Jan. 1966.

[49] Darling Jr. E. M. and Joseph R. D.: "Pattern Recogni-
      tion from Satellite Altitudes", IEEE Trans. on
      System Science and Cybernetics, vol. SSC-4,
      n° 1, pp. 38-47, March 1968.

[50] Das Subrata K.: "A Method of Decision Making in Pat-
      tern Recognition", IEEE Trans. on Computers,
      vol. C-18, n°4, pp. 329-333, April 1969.

[51] Das S. K. and Mohn W. S.: "A Scheme for Speech Process
      ing in Automatic Speaker Verification", IEEE
      Trans. on Audio and Electroacoustics, vol. AU-
      19, n° 1 pp. 32-43, March 1971.

[52] Davenport Jr. W. B. and Root W. L.: "An Introduction to
      the Theory of Random Signals and Noise", New
      York, McGraw-Hill, 1958.

[53] Deutsch E. S.: "On Parallel Operations on Hexagonal Ar
      rays", IEEE Trans. on Comp., vol. C-19, n°
      10, pp. 982-983, October 1970.

[54] Deutsch Sid: "Models of Nervous System", New York,
      John Wiley and Sons, Inc. 1967.

[55]   Dimond T. L.: "Devices for Reading Handwritten Char
        acters", Proc. 1957 Eastern Joint Computer
        Conference, pp. 232-237.

[56]   Dixon R. C. and Boudreau P. E. : "Mathematical Model
        for Pattern Verification", IBM Journal of re
        search and development, vol. 13, n° 6, pp.
        717-721, November 1969.

[57]   Dodwell P. C.: "Visual Pattern Recognition", New York,
        Holt, Rinehart and Winston, 1970.

[58]   Drucker Harris: "Computer Optimization of Recognition
        Networks", IEEE Trans. on Computers, vol.
        C-18, n°10 pp. 918-923, Oct. 1969.

[59]   Duda R.O. and Fossum H.: "Pattern Classification by
        Iteratively Determined Linear and Piecewise
        Linear Discriminant Functions", IEEE Trans.
        on Electr. Comp., vol. EC-15, n° 2, pp. 220-
        232, April 1966.

[60]   Eden M.: "Handwriting and Pattern Recognition", IRE
        Trans. on Inf. Theory, vol. IT-8, n° 2, pp.
        160-166, Feb. 1962.

[61]   Fawe A.L.: "Interpretation of Infinitely Clipped Speech
        Properties", IEEE Trans. on Audio and Elec
        troacoustics, vol. AU-14, n° 4, pp. 178-183,
        Dec. 1966.

[62]   Feigenbaum E. A. and Feldman J., editors: "Computers
        and Thought", New York McGraw-Hill, 1953.

[63]   Figueiredo R. J. P.: "Convergent Algorithms for Pattern
        Recognition in Nonlinearly Evolving Nonsta-
        tionary Environment", Proc. IEEE, vol. 56,
        n° 2, pp. 188-189, Feb. 1968. (Proceedings
        Letter).

[64] Fine T. and Johnson N.: "On the Estimation of the Mean
            of a Random Process", Proc. IEEE vol. 53,
            n° 2, pp. 187-188, Feb. 1965 (Correspondence)

[65] Firschein O. and Fischler M.: "Automatic Subclass De
            termination for Pattern Recognition Applica-
            tions", IEEE Trans. on Electr. Comp., vol.
            EC-12, n°2, pp. 137-141, April 1963. (Cor-
            respondence).

[66] Fisher R.A.: "Statistical Methods for Research Work
            ers", 13th Edition, New York, Hafner, 1963.

[67] Fralick S.C.: "Learning to Recognize Patterns With-
            out a Teacher", IEEE Trans. on Inf. Theory,
            vol. IT-13, n°1, pp. 57-64, Jan. 1967.

[68] Freeman H. and Garder L.: "Apictorial Jigsaw Puzzles:
            The Computer Solution of a Problem in Pattern
            Recognition", IEEE Trans. on Electr. Comp.,
            vol. EC-13, n°2, pp. 118-127, April 1964.

[69] Fu K.S., Min P.I. and Li T.I.: "Feature Selection in
            Pattern Recognition", IEEE Trans. on System
            Science and Cybernetics, vol. SSC-6, n° 1,
            pp. 33-39, January 1970.

[70] Fu K.S., Landgrebe D.A. and Phillips T.L.: "Informa
            tion Processing of Remotely Sensed Agricul-
            tural Data", Proc. IEEE, vol. 57 n° 4, pp.
            639-653, April 1969.

[71] Fukushima Kunihiko: "Visual Feature Extraction by a
            Multilayered Network of Analog Threshold El
            ements", IEEE Trans. on System Science and
            Cybernetics, vol. SCC-5, n° 4, October 1969
            (Notice corrections in vol SCC-6, n°3 p. 239).

[72] Genchi H., Mori K.I., Watanabe S. and Katsuragi S.:
            "Recognition of Handwritten Numerical Char
            acters for Automatic Letter Sorting", Proc.
            IEEE, vol. 56, n°8, pp. 1292-1301 Aug. 1968.

[73] Gitman I. and Levine M. D.: "An Algorithm for Detecting Unimodal Fuzzy Sets and its Application as a Clustering Technique", IEEE Trans. on Computers, vol. C-19 n°7, pp. 583-593, July 1970.

[74] Glucksman H.: "A Parapropagation Pattern Classifier" IEEE Trans. on Electr. Comp., vol. EC-14, n° 3, pp. 434-443, June 1965.

[75] Glucksman H.: "On the Improvement of a Linear Separation by Extending the Adaptive Process with a Stricter Criterion", IEEE Trans. on Electr. Comp., vol. EC-15, n° 6, pp. 941-944, Dec. 1966. (Short note).

[76] Golay M. J. E.: "Hexagonal Parallel Pattern Transformations", IEEE Trans. on Computers, vol. C-18, n° 8, pp. 733-740, August 1969.

[77] Gold B.: "Machine Recognition of Hand-Sent Morse Code" IRE Trans. on Inf. Theory vol. IT-5, n° 2, pp. 17-24, March 1959.

[78] Golshan N. and Hsu C. C.: "A Recognition Algorithm for Handprinted Arabic Numerals", IEEE Trans. on System Science and Cybernetics, vol. SSC-6 n°3, pp. 246-250 July 1970. (Correspondence).

[79] Good I. J.: "The Loss of Information due to Clipping a Waveform", Information and Control, vol. 10, n° 2, pp. 220-222, Feb. 1967.

[80] Good I. J.: "A Five-Year Plan for Automatic Chess" in Dale and Michie 1968, pp. 89-118.

[81] Greenblatt R., Eastlake D. and Crocker S.: "The Greenblatt Chess Program", 1967 Fall Joint Computer Conference, AFIPS Proc., vol. 31, Washington D. C., Thompson, pp. 801-810.

[82]   Groner G.F., Heafer J.F. and Robinson T.W.: "On-
       Line Computer Classification of Handprinted
       Chinese Characters as a Translation Aid",
       IEEE Transl. on Electr. Comp., vol. EC-16
       n°6, pp. 856-860, December 1967. (Correspon
       dence).

[83]   Hajek and Sidak: "Theory of Rank Tests", New York,
       Academic Press Inc., 1967.

[84]   Hald A.: "Statistical Theory with Engineering Applica-
       tions", New York John Wiley and Sons, Fifth
       Printing 1962.

[85]   Hamming R.W.: "On the Distribution of Numbers" BSTJ,
       vol. 49, n° 8, pp. 1609-1625, October 1970.

[86]   Haralick R.M. and Kelly G.L.: "Pattern Recognition
       with Measurement Space and Spatial Cluster-
       ing for Multiple Images", Proc. IEEE, vol.
       57 n° 4, pp. 654-665, April 1969.

[87]   Harmon L.P. and Lewis E.R.: "Neural Modelling",
       Physiological Rev. vol. 46, pp. 513-592, 1966.

[88]   Harmuth H.F.: "A Generalized Concept of Frequency
       and Some Applications", IEEE Trans. on Inf.
       Theory vol. IT-14, n°3 pp. 375-382, May 1968.

[89]   Hart P.E.: "The Condensed Nearest Neighbor Rule",
       IEEE Trans. on Inf. Theory, vol. IT-14, n°
       3, pp. 515-516, May 1968 (Correspondence).

[90]   Hellman Martin E.: "The Nearest Neighbor Classifica
       tion Rule with a Reject Option", IEEE Trans.
       on System Science and Cybernetics, vol. SSC-
       6, n°3, pp. 179-185, July 1970.

[91]   Hellwarth G.A. and Jones G.D.: "Automatic Condition
       ing of Speech Signals", IEEE Trans. on Audio
       and Electroacoustics, vol. Au-16, n° 2, pp.
       169-179, June 1968.

[92] Henderson T.L. and Lainiotis D.C.: "Application of
      State-Variable Techniques to Optimal Feature
      Extraction-Multichannel Analog Data", IEEE
      Trans. on Inf. Theory, vol. IT-16 n° 4, pp.
      396-406, 1970.

[93] Highleyman W.H.: "Design and Analysis of Pattern Rec
      ognition Experiments", Bell System Technical
      Journal, vol. 41, pp. 723-744, March 1962.

[94] Highleyman W.H.: "Linear Decision Functions with Ap-
      plication to Pattern Recognition", Proc. IRE,
      50, n° 6, pp. 1501-1514, June 1962.

[95] Highleyman W.H.: "Data for Character Recognition
      Studies", IEEE Trans. on Electr. Comp., vol.
      EC-12, n° 2, pp. 135-136, April 1963 (Cor-
      respondence).

[96] Hilborn Jr. C.G. and Lainiotis D.G.: "Optimal Unsuper
      vised Learning Multicategory Dependent Hy-
      potheses Pattern Recognition", IEEE Trans.
      on Information Theory, vol. IT-14, n° 3, pp.
      468-470, May 1968.

[97] Hilborn Jr. C.G. and Lainiotis: "Recursive Computa-
      tions for the Optimal Tracking of Time-Vary-
      ing Parameters", IEEE Trans. on Informa-
      tion Theory, vol. IT-14, n° 3, pp. 514-515,
      May 1968.

[98] Hilborn Jr. C.G. and Lainiotis D.G.: "Unsupervised
      Learning Minimum Risk Pattern Classification
      for Dependent Hypotheses", IEEE Trans. on
      System Science and Cybernetics, vol. SSC-5,
      n° 2, pp. 109-115, April 1969.

[99] Ho Yu-Chi and Agrawala A.K.: "On Pattern Classifica
      tion Algorithms Introduction and Survey ",
      Proc. IEEE, vol. 56, n° 12, pp. 2101-2114,
      December 1968.

[100] Ho Y. C. and Kashyap: "An Algorithm for Linear In-
       equalities and its Applications", IEEE Trans.
       on Electr. Comp., vol. EC-14, n° 5,   pp.
       683-688, Oct. 1965.

[101] Hodges D. A.: "Large-Capacity Semiconductor Memo-
       ry", Proc. IEEE, vol. 56, n° 7, pp.   1148 -
       1162, July 1968.

[102] Hoffman R. L. and Moe M. L.: "Sequential Algorithm
       for the Design of Piecewise Linear Classifiers"
       IEEE Trans. on Systems Science and Cyber
       netics, vol. SSC-5, n° 2, pp. 166-168, April
       1969.

[103] Hoffman W. C.: "The Lie Algebra of Visual Perception"
       Boeing Scientific Research Laboratories, Di-
       82-0432, Mathematical Note n°408, April 1965.

[104] Horwitz L. P. and Shelton Jr. G. L.: "Pattern Recogni
       tion Using Autocorrelation", Proc. IRE, vol.
       49, n° 1, pp. 175-185, Jan. 1961.

[105] Howard R. A.: "Value of Information Lotteries", IEEE
       Trans. on Systems Science and Cybernetics,
       vol. SSC-3, n° 1, pp. 54-60, June 1967.

[106] Hu Ming-Kuei: "Visual Pattern Recognition by Moment
       Invariants" IRE Trans. on Information Theory,
       vol. IT-8, n° 2, pp. 179-187, Feb. 1962.

[107] Hughes G. F.: "On the Mean Accuracy of Statistical Pat
       tern Recognizers" IEEE Trans. on Inf. Theory,
       vol. IT-14, n° 1, pp. 55-63, Jan. 1968; for a
       discussion see vol. IT-15, pp. 420-425.

[108] Idzik Jr. S. W. and Misewicz L. M.: "Microminiature
       Programmable Logic Elements", ASTIA n °
       AD 601086, April 1964.

[109] Ingram M. and Preston K. : "Automatic Analysis of Blood Cells", Scientific American pp. 72-82, November 1970.

[110] Irani K. B. : "A Finite-Memory Adaptive Pattern Recognizer", IEEE Trans. on Systems Science and Cybernetics, vol. SSC-4 n°1, pp. 2-11, March 1968.

[111] Ishida H. and Stewart R. M. Jr.: "A Learning Network Using Adaptive Threshold Elements", IEEE Trans. on Electr. Comp., vol. EC-14, n° 6, pp. 481-485, June 1965, (Short note).

[112] Ito T.: "A Note on a General Expansion of Functions of Binary Variables", Information and Control, vol. 12, n° 3, pp. 206-211, March 1968.

[113] Jaynes E. T.: "Prior Probabilities", IEEE System Science and Cybernetics, vol. SSC-4, n° 3, pp. 227-241, Sept. 1968.

[114] Kamenstky L. A. and Liu C. N.: "A Theoretical and Experimental Study of a Model for Pattern Recognition" in Tou and Wilcox 1964, pp. 194-218.

[115] Kanal L. N., editor: "Pattern Recognition", Washington D. C., Thompson Book Comp., 1968.

[116] Kashyap R. L. and Blaydon C. C.: "Recovery of Functions from Noisy Measurements Taken of Randomly Selected Points and its Application to Pattern Classification", Proc. IEEE vol. 54 n° 8, pp. 1127-1129, Aug. 1966 (Correspondence).

[117] Kashyap R. L. and Blaydon C. C.: "Estimation of Probability Density and Distribution Functions ", IEEE Trans. on Inf. Theory, vol. 1T-14, n ° 4, pp. 549-556, July 1968.

[118] Kazmierczak H. and Steinbuch K.: "Adaptive Systems
       in Pattern Recognition" IEEE Trans. on Electr.
       Comp. vol. EC-12 n°6, pp. 822-835 Dec. 1963.

[119] Keehn D. G.: "A Note on Learning for Gaussian Prop-
       erties", IEEE Trans. on Inf. Theory vol. IT-
       -11, n° 1, pp. 126-132, Jan. 1965.

[120] Kersta L.G.: "Voiceprint Identification", Nature vol.
       196, n° 4861, pp. 1253-1257, Dec. 1962; also
       available as Bell Telephone System Monograph
       n° 4485.

[121] Klir Jiri and Valach M.: "Cybernetic Modelling" Iliffe
       Books Ltd., Dorset House Stanford Street ,
       London S. E. 1, 1967.

[122] Knoke P. J. and Wiley R. G.: "A Linguistic Approach to
       Mechanical Pattern Recognition" First Annual
       IEEE Computer Conference, Sept. 1967.

[123] Knoll A. L.: "Experiments with" Characteristic Loci"
       for Recognition of Handprinted Character ",
       IEEE Trans. on Computers, vol. C-18, n°
       4, pp. 366-372, April 1969.

[124] Koford I. S. and Groner G. F.: "The Use of an Adaptive
       Threshold Element to Design a Linear Optimal
       Classifier" IEEE Trans. on Information The-
       ory, vol. 12, n° 1, pp. 42-50, Jan. 1966.

[125] Ku H. H. and Kuliback S.: "Approximating Discrete
       Probability Distributions", IEEE Trans. on
       Information Theory, vol. IT-15, n° 4, July
       1969.

[126] Kulikowski C. A.: "Pattern Recognition Approach to
       Medical Diagnosis", IEEE Trans. on Systems
       and Cybernetics, vol. SSC-6, n° 3, July 1970.

[127]   Kullbach S.: "Information Theory and Statistics" New
        York, John Wiley and Sons, Inc. 1959.

[128]   Kunov H.: "On Recovery in a Certain Class of Neuris
        tors", Proc. IEEE, vol. 55, n° 3, pp. 428-
        429, March 1967 (Proceedings Letter).

[129]   Lainiotis D. G.: "Sequential Structure and Parameter-
        adaptive Pattern Reocgnition - Part I: Supervis
        ed Learning", IEEE Trans. on Information The
        ory vol. IT-16, n° 5, pp. 548-556, Sept. 1970 .

[130]   Laski J.: "On the Probability Density Estimation",
        IEEE, vol. 56, n° 5, pp. 866-867, May 1968
        (Letter).

[131]   Ledley R. S.: "Automatic Pattern Recognition for Clin
        ical Medicine", Proceedings of the IEEE, vol.
        57, n° 11, pp. 2017-2035, Nov. 1969.

[132]   Lehmann E. L.: "Testing Statistical Hypotheses", New
        York, John Wiley and Sons, Inc. 1959.

[133]   Lendaris G. G. and Stanley G. L.: "Diffraction-Pattern
        Sampling for Automatic Pattern Recognition "
        Proc. IEEE, vol. 58, n° 2, pp. 198-216, Feb.
        1970.

[134]   Levine M. D.: "Feature Extraction: A Survey", Proc.
        IEEE, vol. 57, n° 8, pp. 1391-1407, August
        1969.

[135]   Lewis P. M. II: "Approximating Probability Distribu-
        tions to Reduce Storage Requirements", Infor
        mation and Control, vol. 2, pp. 214-225, Feb.
        1959.

[136]   Lewis P. M. II: "The Characteristic Selection Problem
        in Recognition Systems", IRE Trans. on Inf.
        Theory, vol. IT-8, n°2, pp. 171-179, Febr.
        1962.

[137] Lewis P.M. II and Coates C.L. : "Threshold Logic",
       New York, John Wiley and Sons 1967.

[138] Licklider J.C.R.: "Intelligibility of Amplitude-Dichot
       omized, Time-Quantized Speech Waves" Journ.
       of the Acoustical Society of America, vol. 22,
       n° 6, pp. 820-823, Nov. 1950.

[139] Lin T.T. and Yau S.S. : "Bayesian Approach to the Op-
       timization of Adaptive Systems", IEEE Trans.
       on Systems Science and Cybernetics, vol. SSC-
       -3, n° 2, pp. 77-85, Nov. 1967.

[140] Liu C.H. and Shelton G.L.Jr.: "An Experimental In-
       vestigation of a Mixed-Font Print Recognition
       System", IEEE Trans. on Electr. Comp. vol.,
       EC-15, n° 4, pp. 916-925, Dec. 1966.

[141] Marill T. and Green D.M. : "Statistical Recognition
       Functions and the Design of Pattern Recogni-
       zers", IRE Trans. on Electr. Comp., vol.
       EC-9, n° 4, pp. 472-477, Dec. 1960.

[142] Marill T. and Green D.M.: "The Effectiveness of Re
       ceptors in Recognition Systems", IEEE Trans.
       on Inf. Theory, vol. IT-9, n° 1, pp. 11-17,
       Jan. 1963.

[143] Martin Francis F.: "Computer Modelling and Simula
       tion", New York, John Wiley and Sons, Inc.
       1968.

[144] Mattson R.L. and Dammann J.E.: "A Technique for De
       termining and Coding Subclasses in Pattern Rec
       ognition Problems", IBM Journal of R and D,
       vol. 9, n° 4, pp. 294-302, July 1965.

[145] Meisel W.S.: "Potential Functions in Mathematical Pat
       tern Recognition" IEEE Trans. on Computers,
       vol. C-18, n° 10, pp. 911-918 October 1969.

[146] Mendel I. M. and Fu K. S., editors: "Adoptive Learning
         and Pattern Recognition Systems", New York
         Academic Press, 1970.

[147] Mengert P. H.: "Solution of Linear Inequalities" IEEE
         Trans. on Computers, vol. C-19, n° 2, pp.
         124-131, Febr. 1970.

[148] Mermelstein P. and Eden M.: "Experiments on Com-
         puter Recognition of Connected Handwritten
         Words", Information and Control, vol. 7, n°
         2, pp. 255-270, June 1964.

[149] Middleton D.: "An Introduction to Statistical Communi
         cation Theory, New York, McGraw-Hill, 1960.

[150] Miller R. G.: "An Application of Multiple Discriminant
         Analysis to the Probabilistic Prediction of Me
         teorological Conditions Affecting Operational
         Decisions", Travellers Reserach Center, Inc.
         Hartford, Conn. TRCM-4, March 1961.

[151] Minsky M.: "Steps Toward Artificial Intelligence"Proc.
         IRE, vol. 49, n° 1, pp. 8-30, Jan. 1961.

[152] Minsky M.: "A Selected Descriptor-Indexed Bibliogra-
         phy to the Literature on Artificial Intelligence'
         in Feigenbaum and Feldman 1963, pp. 152-523.

[153] Minsky M. and Papert S.: "Perceptrons", Cambridge,
         Mass. 02142, The MIT Press, 1969.

[154] Munson J. H.: "Experiments in the Recognition of Hand
         printed Text: Part I - Character Recognition",
         Proc. 1968 Fall Joint Computer Conference,
         vol. 33, pp. 1125-1138. (Contains reviews of
         more than 40 papers on the recognition of hand
         printed characters).

[155] Munson J.H., Duda R.O. and Hart P.E.: "Experiments with Highleymans Data" IEEE Trans. on Electr. Comp., vol. EC-17, n° 4, pp. 399-401, April 1968 (Correspondence).

[156] Muses C.A., editor: "Aspects of the Theory of Artifi cial Intelligence", New York, Plenum Press, 1962.

[157] Nagy G.: "State of the Art in Pattern Recognition", Proc. IEEE, vol. 56, n° 5, pp. 836-862, May 1968.

[158] Nagy G.: "Feature Extraction on Binary Patterns", IEEE Trans. on Systems Science and Cybernetics, vol. SCC-5 n° 4, pp. 273-278 Oct. 1969.

[159] Nelson G.D. and Levy D.M.: "A Dynamic Programm ing Approach to the Selection of Pattern Features", IEEE Trans. on Systems Science and Cybernetics, vol. SSC-4, n° 2, pp. 145-152, July 1968.

[160] Nelson G.D. and Levy D.M.: "Selection of Pattern Fea tures by Mathematical Programming Algorithms", IEEE Trans. on Systems Science and Cybernetics, vol. SS-6, n° 2, pp. 145-151 Jan. 1970.

[161] Newell A., Shaw J.C. and Simon H.A.: "Chess-Playing Programs and the Problem of Complexity" in Feigenbaum and Feldman 1963, pp. 39-70.

[162] Newell A. and Simon H.A.: "An Example of Human Chess Play in the Light of Chess Playing Pro grams" in "Progress in Biocybernetics vol. 2" Norbart Wiener and J.P. Schade editors New York, American Elsevier Publishing Co. Inc. 1965.

[163]   Nielsen K. A., M.E.E. Thesis, Electronics Lab., Tech.
        University of Denmark, Lyngby, 1969.

[164]   Nilsson Nils J.: "Learning Machines", New York, Mc-
        Graw-Hill, 1965.

[165]   Oppenheim A. V., Schafer R. W. and Stockbaum Jr. T. G.
        "Nonlinear Filtering of Multiplied and Control-
        led Signals", Proc. IEEE, vol. 56, n°8, pp.
        1264-1291, August 1968.

[166]   Page J.: "Recognition of Patterns in Jet Engine Vibra
        tion Signals", Digest of the First Annaul IEEE
        Computer Conference (Sep. 6-8, 1967), New
        York, IEEE Publications n°16 C 51, pp. 102-
        105.

[167]   Paramentier R. D.: "Neuristor Analysis Techniques
        for Nonlinear Distributed Electronic Systems"
        Proc. IEEE vol. 58, n°11, pp. 1829-1837,
        November 1970.

[168]   Patrick E. A. and Hancock J. C.: "Nonsupervised Se-
        quential Classification and Recognition of Pat
        terns", IEEE Trans. on Inf. Theory, vol. IT-
        n° 3, pp. 362-372, July 1966.

[169]   Peterson D. W. and Mattson R. L.: "A Method of Find
        ing Linear Discriminant Functions for a Class
        of Performance Criteria", IEEE Trans. on
        Inf. Theory, vol. IT-12, n°3, pp. 380-387,
        July 1966.

[170]   Prewitt J. M., Mayall B. H. and Mendelsohn M. L.:
        "Pictorial Data Processing Methods in Micro
        scopy", Proc. Soc. Photographic Instrumenta
        tion Engrs. (Boston, Mass), June 1966.

[171]   Raviv J.: "Decision Making in Markov Chains Applied
        to the Problem of Pattern Recognition", IEEE
        Trans. on Inf. Theory vol. IT-13, n°4, pp. 536
        -551, Oct. 1967.

[172] Robbins H. and Monro S.: "A Stochastic Approximation Method", Ann. Math. Stat. vol. 22, pp. 400-407, 1951.

[173] Rosenblatt F.: "Perceptron Experiments", Proc. IRE, vol. 48 n°3, pp. 301-308, March 1960.

[174] Rosenblatt F.: "Principles of Neurodynamics", Washington D.C., Spartan, 1962.

[175] Samuel A.L.: "Some Studies in Machine Learning Using the Game of Checkers" in Feigenbaum and Feldman 1963, pp. 71-105.

[176] Samuel A.L.: "Some Studies in Machine Learning Using the Game of Checkers, II Recent Progress" IBM Journal of R and D, vol. 11, pp. 601-617, Nov. 1967.

[177] Sebestyen George S.: "Decision-Making Processes in Pattern Recognition", New York, The Macmillan Company, 1962.

[178] Sebestyen G. and Edie J.: "An Algorithm for Non-Parametric Pattern Recognition", IEEE Trans. on Electr. Comp., vol. EC-15, n° 6, pp. 908-915, Dec. 1966.

[179] Selfridge O.G.: "Pandemonium: A Paradigm for Learning" in "Mechanization of Thought Processes", London, Her Majesty's Stationary Office, 1959, pp. 513-526; reprinted in Uhr 1966, pp. 339-348.

[180] Selfridge O.G.: "Keynote: Some Notes on the Technology of Recognition" in G.L. Fisher et al. (Editors) "Optical Character Recognition", Washington D.C., Spartan 1962.

[181] Selin Ivan: "Detection Theory" Princeton, New Jersey, Princeton University Press, 1965.

[182] Siegel S.: "Non-Parametric Statistics for the Behavioral Sciences", New York, McGraw-Hill, 1956.

[183] Simek J.G. and Tunis C.J.: "Handprinting Input Device for Computer Systems", IEEE Spectrum, vol. 4, n° 7, pp. 72-81, July 1967.

[184] Smith F.W.: "Pattern Classifier Design by Linear Programming", IEEE Trans. on Comp. vol. C-17, n° 4, pp. 367-372, April 1968.

[185] Solomonoff R.I.: "Some Recent Work in Artificial Intelligence", Proc. IEEE, vol. 54, n° 12, pp. 1687-1697, Dec. 1966.

[186] Specht D.F.: "Generation of Polynomial Discriminant Functions for Pattern Recognition" IEEE Trans on Electr. Comp., vol. EC-16, n° 3, pp. 308-319, June 1967.

[187] Spragins J.: "A Note on the Iterative Application of Bayes' Rule", IEEE Trans. on Inf. Theory, IT-11, n°1, pp. 544-549, Oct. 1965.

[188] Spragins J.: "Learning Without a Teacher", IEEE Trans. on Inf. Theory, vol. IT-12, n° 2, pp. 223-230, April 1966.

[189] Steinbuch K. and Piske U.A.W.: "Learning Matrices and Their Applications" IEEE Trans. on Electr. Comp., vol. EC-12, pp. 846-862, Dec. 1963.

[190] Steinbuch K. and Widrow B.: "A Critical Comparison of Two kinds of Adaptive Networks", IEEE Trans. on Electr. Comp., vol. EC-14, n° 5, pp. 737-740, Oct. 1965.

[191] Strakhov N.A. and Kurz L.: "An Upper Bound on the Zero-Crossing Distribution" Bell System Tech. Journal vol. 47 n° 4, pp. 529-547, April 1968.

[192] Tamura S., Higuchi S. and Tanaka K.: "Pattern Clas
       sification Based on Fuzzy Relations", IEEE
       Trans. on Systems, Man and Cybernetics, vol.
       SMC-1, n° 1, pp. 61-66, January 1971.

[193] Tenery G.: "A Pattern Recognition Function of Integral
       Geometry", IEEE Trans. on Military Electron
       ics, vol. MIL-7, nos. 2 and 3 pp. 196-199,
       April-July 1963.

[194] Thamdrup J.E., M.E.E. Thesis, Electronics Lab.,
       Tehcnical University of Denmark, Lyngby ,
       1969.

[195] Thomas R.B. and Kassler M.: "Character Recognition
       in Context", Information and Control, vol. 10
       n° 1, pp. 43-64, Jan. 1967.

[196] Tou J.T., editor: "Computer and Information Sciences-
       II", New York, Academic Press 1967.

[197] Tou J.T. and Wilcox R.H., editors: "Computer and In
       formation Sciences", Washington D. C.,Spartan
       Books Inc., 1964.

[198] Toussaint G.T. and Donaldson R.W.: "Algorithms for
       Recognizing Contour-Traced Handprinted Char
       acters", IEEE Trans. on Computers, vol. C-
       -19, n° 6, pp. 541-546, June 1970.

[199] Tsypkin Ya.Z.: "Self-Learning. What is it? ", IEEE
       Trans. on Automatic Control, vol. AC-13, n°
       6, pp. 608-612, Dec. 1968.

[200] Uhr L. and Vossler C.: "A Pattern Recognition Pro-
       gram that Generates, Evaluates and Adjusts
       Its Own Operators", 1961 Proceedings of the
       Western Joint Computer Conference, pp. 555-
       -569; reprinted in Feigenbaum and Feldman
       1963, pp. 251-268; again reprinted in Uhr 1966,
       pp. 349-364.

[201] Uhr L.: "Pattern Recognition" in "Electronic Informa_
tion Handling", A. Kent and O.E. Taulbee,
editors, Washington D.C., Spartan Books Inc.,
1965, pp. 51-72; reprinted in Uhr 1966, pp.
365-381.

[202] Uhr Leonard, editor: "Pattern Recognition", New York,
John Wiley and Sons, Inc. 1966.

[203] Ullman J.R.: "Experiments with the n-tuple Method of
Pattern Recognition", IEEE Trans. on Comp.,
vol. C-18, n° 12, pp. 1135-1138, Dec.1969.

[204] Unger S.H.: "Pattern Detection and Recognition ",
Proc. IEEE, vol. 47, pp. 1737-1752, October
1959.

[205] Wagner T.G.: "The Rate of Convergence of an Algo_
rithm for Recovering Functions from Noisy
Measurements Taken at Randomly Selected
Points", IEEE Trans. on Systems Science and
Cybernetics, vol. SSC-4 n° 2, July 1968.

[206] Wee W.G.: "A Survey of Pattern Recognition", Paper
presented at the Seventh Symposium on Adaptive Processes, Dec. 16-18, 1968 at UCLA,
California, U.S.A.

[207] Wee W.G. and Fu K.S.: "An Adaptive Procedure for
Multiclass Pattern Classification" IEEE Trans.
on Computers, vol. C-17 n° 2, pp. 178-182,
Feb. 1968.

[208] Widrow B., Groner G.F., Hu M.I.C., Smith F.W.,
Specht D.F. and Talbert L.R.: "Practical Applications for Adaptive Data-Processing Systems", WESCOW 1963, paper n° 11.4.

[209] Wilks S.S.: "Mathematical Statistics" New York, John
Wiley and Sons, Inc. 1962.

[210]   Winder R.O.: "The Status of Threshold Logic", RCA
        Review, vol. 30 n°1, pp. 62-84, March 1969 .

[211]   Wolff A. C.: "The Estimation of the Optimum Linear
        Decision Function with a Sequential Random
        Method", IEEE Trans. on Inf. Theory, vol.
        IT-12, n° 3, pp. 312-315, July 1966.

[212]   Wolverton C.T. and Wagner T.I.: "Asymptotically
        Optimal Discriminant Functions for Pattern
        Classification", IEEE Trans. on Inf. Theory,
        vol. IT-15, n° 2, pp. 258-265, March 1969.

[213]   Yau S.S. and Yang C.C.: "Pattern Recognition by Us
        ing an Associative Memory", IEEE Trans. on
        Electr. Comp., vol. EC-15, n° 6, pp. 941-
        944, Dec. 1966. (Short Note).

[214]   Yau S.S. and Ling T.T.: "On the Upper Bound of the
        Probability of Error of a Linear Pattern Clas
        sifier for Probabilistic Pattern Classes" Proc.
        IEEE, vol. 56, n° 3, pp. 321-322, March
        1968 (Letter).

[215]   Yau S.S. and Schumpert S. M.: "Design of Pattern Clas
        sifiers with Updating Property Using Stochastic
        Approximation Technique", IEEE Trans. on
        Computers, vol. C-17, n° 9, Sept. 1968.

[216]   Young T.Y. and Huggins W.H.: "Computer Analysis
        of Electro-Cardiograms Using a Linear Re-
        gression Technique", IEEE Trans. on Bio-
        Medical Engineering, vol. BME-11, pp. 60-
        67, July 1964.

ADDENDUM

THE Θ-TRANSFORMATION

ABSTRACT

This paper presents the answer to a problem which may have puzzled many of the readers: given a multivariate probability density function and its marginals, how does one find the other multivariate densities which have the same set of marginals? The answer is that all the multivariate densities may be found by applying a particular transformation (called the - transformation) a number of times to the product of the marginals. It thus becomes possible, by hill-climbing, to find the "worst" or the "best" multivariate density which is concomitant with a specified set of marginals. This new technique is important in connection with problems where the designer only has access to the marginals and where he wants to bound functions of the multivariate density. For illustration, the technique is used to solve two hitherto unsolved problems, one in the field of electronic circuit reliability, and one in the field of pattern recognition.

Peter W. Becker

GLOSSARY OF SYMBOLS

a_j The jth coordinate for P_1, or for P in Appendix 2.

$\bar{a}_j$ $x_j = \bar{a}_j$ means that the value of x_j differs from a_j.

b_j The jth coordinate for P_2 is called $\left(a_j + b_j\right)$.

D A metric expressing the "distance" between f and the transformed f_p.

$_kD$ The value of D after $k\,\Theta$-transformations, defined by Equation 2.

^P_0d $\left(^Xf - {}^Xf_p\right)$ at P is called ^P_0d.

^Pd ^Xd at the point P.

^Xd The difference between ^Xf and the transformed ^Xf_p.

^X_0d A difference defined by Equation 3.

^X_kd A difference defined by Equation 1.

E_1 Sum of excess mass; Appendix 2.

F The average pattern classification error probability, Example 2.

$f = f(X)$ The multivariate probability density function for X.

$f_j = f_j(X_j)$ The marginal probability density function for X_j.

f_p The product of the N marginals.

f_k f_p after it has been modified by k Θ-transformations.

^Xf The value of f at the lattice point X.

^Xf_p The value of f_p at the lattice point X.

^X_kf The value of $_kf$ at the lattice point X.

N	Number of measurements, and dimensionality of measurement space.
n	Number of values of X, thereby number of lattice points.
n_i	Number of discrete values of X_i ; Section 2.
P	Lattice point used in the appendices.
P_1, P_2, P_3, P_4	Lattice points used in the θ-transformation.
$_A P, _B P, _C P$	A priori probabilities, Example 2.
Q	Reliability, used in Example 1.
q	All the discrete probabilities mentioned are assumed to be multiples of q.
S_i	The collection of lattice points for which $X_i = a_i$; S_i is used in the appendices .
$S_{12\ldots k}$	The collection of lattice points for which $(X_1, X_2, ..., X_k) = (a_1, a_2, ..., a_k)$.
W	Number of "admissible sub-sequences of joint densities".
X	Value of a set of N parameters, X is illustrated by a lattice point in measurement space; Section 2.
X_i	The ith parameters; Section 2.
z	Number of θ-transformations in the first step of the first sub-sequence, Appendix 2, Step n° 1.
Δ	Distance used in Figure 1.
Φ	Function used in Example 2.
θ	Probability mass moved simultaneously from P_1 to P_3 , and from P_2 to P_4 ; θ is non-negative.

1. Introduction

In this paper we present the answer to the following problem which may have puzzled many of the readers: given a multivariate probability density function and all its marginal probability density functions, how does one find the other multivariate densities which have the same set of marginals? Although work has been done on related problems[1, 2, 3, 4] this question seems to have been ignored. The answer is: all multivariate density functions which have a specified set of marginals are obtained by repeated application of the Θ-transformation to the product of the marginals; the Θ-transformation will be described in a following section.

The practical value of the answer lies in the fact that it now becomes possible to determine what the "best" and the "worst" multivariate densities in some particular sense would be for a specified set of marginals. With knowledge of these extreme multivariate densities the designer can bound variables of interest. This new procedure will later be illustrated with examples from the field of electronic circuit reliability and the field of pattern recognition.

2. Some Preliminary Assumptions

Let it be assumed that we have repeated an experiment a large number of times; two examples of such experiments will be presented in Examples 1 and 2. At the end of each experiment we measure the values of N parameters, X_i, $i = 1, \ldots, N$.

The set of N parameters are referred to as:

$$X = (X_1, \ldots, X_i, \ldots, X_N).$$

We now make a mild assumption: we assume that X_i always takes one of n_i discrete, fixed values; the assumption is satisfied in practice due to the inherent quantization and limited range of results obtained with measuring equipment. Consequently, the experiment (as judged from the parameter values) can have no more than

$$n = n_1 \cdot n_2, \ldots, n_i, \ldots, n_N$$

different outcomes, each of which may be illustrated by a lattice point in the N-dimensional measurement space. With each lattice point is associated the relative frequency of the corresponding result of the experiment. We assume, that the multivariate probability density function for X, $f = f(X)$, can be estimated from these "relative frequencies", an assumption which

is widely used in engineering science $[5, 6]$. We make a second
mild assumption: we assume that each of the n discrete probabilities which constitute f, is a multiple of some small quantity of probability mass, q. The assumption is justified by the
finite accuracy of the measuring and computing devices.

f has N marginal densities. They are called $f_1 = f_1(X_1), \ldots,$
$f_i = f_i(X_i), \ldots, f_N = f_N(X_N)$; f_i consists of n_i discrete $\underline{prob}$
abilities having a sum of unity. The product of the N marginals
is called f_p ; f_p is a multivariate density function and like f
it has the N marginals $f_i, i = 1, \ldots, N$.

In Sections 3, 4 and 5 both X and f are dis -
crete-valued functions; in Section 6 the case where both X and
f are continuous functions will be treated.

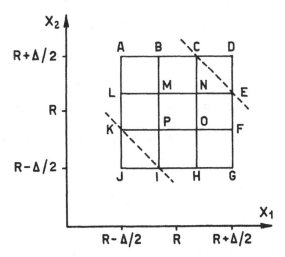

Fig. 1. The bivariate probability density function for the true values of 2 series connected resistors X_1
and X_2 ; the reliability of $(X_1 + X_2)$ is discussed in Example 1.

3. The Θ-Transformation

Consider two lattice points: $P_1 = a_1, \ldots, a_i,$ $\ldots, a_N)$, and $P_2 = (a_1+b_1, \ldots, a_i+b_i, \ldots, a_N+b_N)$; the two values of X are observed with probabilities Θ_1 and Θ_2. Let Θ be a quantity between zero and the smaller of Θ_1 and Θ_2. The Θ-transformation consists of moving Θ units of probability mass from P_1 to a lattice point P_3, while simultaneously moving Θ units of probability mass from P_2 to a lattice point P_4; the locations of P_3 and P_4 are determined as follows: The N coordinates for P_3 are obtained by using some of the P_1 coordinate-values and some of the P_2 coordinate-values; X_1 is a_1 or $(a_1 + b_1)$, X_2 is a_2 or $(a_2+b_2), \ldots, X_i$ is a_i or $(a_i + b_i)$. The above-mentioned coordinate values not used for P_3 are used for P_4 . For illustration, consider the case N=3; P_3 could be (a_1, a_2, a_3+b_3) and P_4 would be $(a_1 + b_1, a_2 + b_2, a_3)$. The Θ-transformation leaves the N marginals unchanged because the probability of a lattice-point having $X_i = a_i$, $f_i(a_i)$, or having $X_i = a_i + b_i$, $f_i(a_i + b_i)$, is unchanged.

The reader should verify for himself that the above statement is true for N = 2. (Figure 2 illustrates the case where N = 3, the case which will be discussed in Appendix 1).

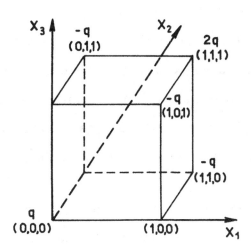

X_3 -q (0,1,1)   X_2   2q (1,1,1)

-q (1,0,1)

-q (1,1,0)

q (0,0,0)   (1,0,0)   X_1

Fig. 2.

A trivariate case.   The q-values apply to the case in
Appendix 1.

4. A Theorem

We are only concerned with joint densities which consist of non-negative discrete probabilities; such joint densities will be called "admissible".

The set of N marginals is said to be "admissible" if all marginals consist of non-negative discrete probabilities; F_p, the product of the N members of an admissible set of marginals, clearly is an admissible joint density. An admissible joint density always has an admissible set of N marginals. An "admissible sequence of joint densities" is a sequence of admissible joint densities, $_1f$, $_2f$, $_3f$, etc., each of which is obtained from its predecessor by one Θ-transformation. When the values of n and q are specified, the number of possible, admissible joint densities obviously is bounded. The theorem can now be stated.

Theorem: Let f be a discrete admissible joint density with the marginals $(f_1, \ldots, f_N)$; then there exists at least one admissible sequence of joint densities which begins with F_p and ends with f, and which is a sequence of finite length.

The importance of the theorem lies in the fact that it insures us against the following unpleasant possibility. One could imagine that to reach an optimum "admissible joint density" by hillclimbing from f_p it might be necessary to pass through nonadmissible joint densities; the theorem tells us that this is not so.

The theorem does not tell the hill-climbing designer how to find the multivariate density which has some particular property in largest measure; the theorem only states that the interesting multivariate density can be generated from f by a finite number of applications of the Θ-transformation, each of which changes one admissible density to another admissible density.

Whenever hill-climbing is used to solve engineering problems, it always becomes necessary to generate a starting point or a "feasible solution"; with our problem the starting point is f_p. It is a distinct advantage in our case, that we get to the starting point so easily. f_p is simply the product of the N marginals; with many other engineering problems just getting to the starting point can be quite a problem.

5. A Proof of the Theorem

We are concerned with two multivariate densities, f_p and f, and their values ^Xf_p and ^Xf at the lattice point X. f_p is the starting point, f is the fixed goal for our transformations. The first Θ-transformation changes f_p to $_1f$, the second Θ-transformation changes $_1f$ to $_2f$, the k'th Θ-transformation changes $_{(k-1)}f$ to $_kf$, etc. According to the theorem, f_p can be transformed to f through a finite number of Θ-transformations. To ascertain how different $_kf$ is from f we introduce the "distance function", $_kD$. The metric $_kD$ is defined as the sum over all n lattice points of the numerical values of ^X_kd; ^X_kd and $_kD$ are defined by Equations 1 and 2.

$$^X_kd = {}^Xf - {}^X_kf \qquad (1)$$

$$_kD = \Sigma \left| {}^X_kd \right| \qquad (2)$$

all lattice points

For $k = 0$ we have the following two expressions:

$$^X_0d = {}^Xf - {}^Xf_p \qquad (3)$$

$$_0D = \Sigma \left| {}^X_0d \right| \qquad (4)$$

all lattice points

The form of the metric is not critical; ${}^x d$ could equally well have been defined as $\left({}^x_k f - {}^x_k f\right)^{y}$, with y being some fixed, positive, even integer. The main thing is, that **D** is the sum of n terms, one for each lattice point, where the value of each term decreases monotonically with decreasing value of $\left|{}^x_k d\right|$; also it is desirable that the value of each term becomes zero for ${}^x_k d = 0$, because it ensures that the terminal value of **D** is zero. At each lattice point, X, ${}^x_k d$ is the goal for our net tran<u>s</u> fer of probability mass to the point. At the lattice points where ${}^x_k d$ is positive, we will say that probability mass is needed; at points where ${}^x_k d$ is negative, we will say that probability mass should be removed. <u>Probability mass which should be re- moved is called excess probability mass; it is only found at lat</u> tice points where ${}^x_k d$ is negative. It will be standard procedure in our manipulations never to move probability mass away from lattice points with positive ${}^x_k d$ -values, and <u>at the most to move</u> $\left(-{}^x_k d\right)$ <u>away from lattice points with negative</u> ${}^x_k d$ <u>-values.</u> This procedure has two advantages. (1) We are assured against ge<u>n</u> erating a "not admissible density" from an "admissible density" because we always leave at least ${}^x f$ units of probability mass at a lattice point. (2) Also, we are assured that the value of **D** does not increase: the reason for this will now be presented. If in Θ-transformation $N^\circ (k+1)$ we move Θ units of excess prob<u>b</u> ability mass from a point, P_1 , to a point, P_3 , where there already is too much probability mass, (${}^{P_3}_k d$ is <u>negative</u>) the positive change

$$_{(k+1)}^{P_1}d - _{k}^{P_1}d = \Theta$$

is equaled by the negative change

$$_{(k+1)}^{P_3}d - _{k}^{P_3}d = -\Theta$$

when $_{(k+1)}D$ is computed using Equation 2; no change results in the value of D, $_{(k+1)}D = _{k}D$: If excess probability mass is needed at P_3 ($_{k}^{P_3}d$ is positive), however, the value of D will decrease by an amount between 0 and 2Θ. What was just said about the shift of probability mass from P_1 to P_3 also holds true for the simultaneous shift of Θ units of probability mass from P_2 to P_4 ; the decrease in D-value due to one Θ-transformation is between 0 and 4Θ.

We are consequently assured of the following relationship:

$$_{0}D \geq _{1}D \geq D_2 \geq \ldots \geq _{k}D \geq _{(k+1)}D. \tag{5}$$

After these introductory remarks we can now present the proof of the theorem.

In Appendix 2 a particular procedure for generating "an admissible sub-sequence of joint densities" is described, the procedure only involves the moving of excess probability mass from lattice points with negative $_{k}^{X}d$ -values. At the end of Appendix 2 it is explained why the procedure has two interesting features. (1) The procedure insures that the length of the sub-sequence (meaning the number of Θ-transformations performed) is bounded by (a function of $n_1, n_2, \ldots, n_N$) (2) At the end of the sub-sequence the value of the "distance measure", D, is guaranteed to have decreased by at least 2q; the value of D will probably have decreased by much more. By repeating the procedure a finite number of times, **W,** the value of D can be reduced to zero. W is finite for the following reason. Equa tion 4 shows that the "distance" $_0D$ between f and f_p can be no more than two units of probability mass and that $_0D$ must be a multiple of q. As mentioned earlier in this paragraph, the value of D is reduced by at least 2q after each "admissible sub-sequence". The number, W, of "admissible sub-sequences" can consequently be no more than $2/(2q) = 1/q$.

The string of W "admissible sub-sequences of joint densities" (each of which involves a finite number of Θ-transformations) constitutes one possible "admissible se-quence of joint densities beginning with f_p and ending with f", the existence of which was promised in the theorem. This completes the proof.

For the sake of brevity, "probability mass" and "lattice points" will in the following often be referred to as "mass" and "points".

6. The Case of Continuous Densities

If f is a continuous multivariate density function, rather than a discrete multivariate density function, as
hitherto assumed, the situation changes somewhat. As before,
we have the N marginals and their product f_p , all of which are
continuous densities. Also, we can redefine $_k^X d$ as the numerical value of the density-difference at the point X after the kth
Θ-transformation; how a Θ transformation may be performed
will be described shortly.

$$_k^X d = {}_{}^X f - {}_k^X f \ .$$

We can redefine $_k D$ as the integral of $\left| {}_k^X d \right|$ over the N-dimen-
sional volume V of interest.

$$_k D = \int_V \left| {}_k^X d \right| (dV)$$

We can also reduce the value of D in a systematic way as before; e. g., we could do the following. Four points, P_1 , P_2 , P_3 ,
and P_4 are selected as described in the beginning of Section 3.
At each point, and in the same manner, we locate identical orthotopes (meaning N-dimensional boxes with sides vertical to
the axis) $O_1, O_2, O_3,$ and O_4 . We redefine Θ as some continuous
function defined over the orthotope. If we now substitute $_k f$
with $\left({}_k f - \Theta \right)$ inside O_1 and O_2 , and with $\left({}_k f + \Theta \right)$ inside O_3 and

it is seen that the marginals are retained. If $\left({}_{K}f - \Theta\right)$ inside O_1 and O_2, and $\left({}_{K}f + \Theta\right)$ inside O_3 and O_4, all are non negative functions, the Θ-transformation is admissible. The main difference between the discrete and the continuous case is that in the continuous case there is no assurance that a finite number of Θ-transformations can force the value of D to be identically equal to zero. Examples 1 and 2 illustrate two cases where continuous densities are being modified by a sequence of admissible Θ-transformations.

Example 1

The Reliability of Two Resistors Connected in Series

Consider the case where we have two resistors connected in series. The resistors have the true values x_1 and x_2 ; they both have the nominal value R. Both x_1 and x_2 have flat density functions with ranges from $R - \Delta/2$ to $R + \Delta/2$; $f_1 = f_1(x_1) = f_2 = f_2(x_2) = 1/\Delta$ within the range and zero outside. We will assume that the knowledge about f_1 and f_2 has been obtained by experiments where the resistance of randomly selected resistors was measured. In the case at hand $N = 2, x = (x_1, x_2), f$ obviously is zero outside the square ADGJ, Figure 1. Let it for some particular application be desired that the value of $(x_1 + x_2)$ falls in the open interval from $(2R - 2\Delta/3)$ to $(2R + 2\Delta/3)$. The probability with which this happens we will call the reliability, Q. When a (x_1, x_2)-combination satisfies the constraints, the corresponding point falls within the hexagon ACEGIK. The question to be answered in this example is: which density among all bivariate density functions with marginals f_1 and f_2 will minimize the reliability Q? Question of this kind can only be answered by using the Θ-transformation. We start with the density $f_p = f_1 \cdot f_2$; for this function the bivariate density is $1/\Delta^2$ over the square ADGJ and $Q = 8/9$. We now apply a series of Θ-transformations to f_p , each of which reduces Q; it is then hoped for that the hill-climbing will lead to a Q-value which is globally minimum.

The procedure is the one described in Section 6, "The Case of the Continuous Densities". In the first 3 transformations the orthotopes are square with sidelength $\Delta/3$; we use $\Theta = 1/\Delta^2$, and this makes $f_p + \Theta = 2/\Delta^2$ and $f_p - \Theta = 0$.

1) The probability mass in LMPK is moved to MNOP, and at the same time the mass in POHI is moved to KPIJ, Q is thereby reduced to 5/6.

2) The mass in BCNM is moved to CDEN, and at the same time the mass in NEFO is moved to MNOP. Q is now 7/9.

3) The mass in ABML is moved to CDEN, and at the same time the mass in OFGH is moved to KPIJ. The density $_{,3}f$, is $3/\Delta^2$ over the squares CDEN, MNOP and KPIJ, and zero over the 6 other little squares. Q is now 2/3.

We could obtain the same value of Q with many other densities; e.g., if all the mass is evenly distributed on the diagonal ID, Q will also equal 2/3. Densities with Q near 2/3 call for a strong positive correlation between x_1 and x_2 ; such correlations may very well be realistic for component values from integrated electronic circuits. If one is willing to accept pathological densities, one can proceed from the $_3f$ distribution above and with an infinite sequence of Θ-transformations locate the mass in KPIJ on the diagonal KI, and the mass in CDEN on the diagonal CE. According to the specifications the points on CE and KI illustrate non-permissible values of $(x_1 + x_2)$; this means that $Q = 1/3$

for the density. The interesting thing about the density is that it represents the global minimum for Q, which may be verified by inspection.

In this example we made use of the idea that constraints on a circuit (e.g., two resistors in series) may be illustrated by a region in (X_1, X_2)-space (e.g., the hexagon ACEGIK). The point (x_1, x_2), that illustrates the actual circuit should fall within the region if and only if the circuit works satisfactorily. This idea can also be used with complex circuits [8, 9]; the use of Θ-transformations to assess the minimum reliability has therefore more potential than what transpires from the two resistor example.

Example 2
The Largest Average Pattern-Classification Error-Probability.

In the field of Pattern Recognition it is important to determine the magnitude of the average probability of pattern misclassification, F.

Also, it is important to determine what the largest possible value of F is; the value is called F_{Max}. Questions of this kind can only be answered by using the Θ-transformation to minimize F. We will now through an example outline the pro cedure for finding F_{Max} .

Let it be assumed that we are concerned with members of the three pattern classes, Class A, Class B and Class C; each member is assumed to belong to one and only one of the classes. We select representative members from the three classes and measure their values, x_1 , of the attribute X_1 ; from the results we obtain $_Af_1 = _Af_1(x_1)$, $_Bf_1 = _Bf_1(x_1)$, and $_Cf_1 = _Cf_1(x_1)$, the probability density functions for members of Class A, Class B, and Class C. We then measure the values, x_2 , of the attribute X_2 for a second set of representative members and obtain the densities $_Af_2 = _Af_2(x_2)$, $_Bf_2 = _Bf_2(x_2)$, and $_Cf_2 = _Cf_2(x_2)$.
The members of Class A have a bivariate density $_Af = _Af(x_1, x_2)$ with marginals $_Af_1$ and $_Af_2$; we know $_Af_1$ and $_Af_2$ but not $_Af$.
Likewise we do not know the bivariate density, $_Bf = _Bf(x_1, x_2)$, for Class B members, or $_Cf = _Cf(x_1, x_2)$ for Class C members; only the marginals are known. Our experiment consisted in this case of selecting a representative member from a pattern class and measuring its value of an attribute.

We are concerned with the following classification problem. An unlabelled pattern is known to be a member of either Class A, Class B or Class C, its values of the attributes X_1 and X_2 being $(x_1, x_2) = (s_1, s_2)$. The a priori probability of class membership is assumed to be known and remain fixed for each of the three classes; the a priori probabilities are called $_Ap$, $_Bp$ and $_Cp$, their sum is unity. The problem is now to determine the

densities, $_Af$, $_Bf$ and $_cf$, which will result in the <u>largest aver-</u>
<u>age probability of classification error</u>; the densities should be
concomitant with the three pairs of marginals. To the best of
the authors knowledge this fundamental problem has not been
solved before.

In realistic cases $_Af_1$ and $_Af_2$ both are identically zero outside
some finite range; this means that if a member of Class A is
illustrated by its two attribute values (x_1, x_2), the correspond
ing point in the (X_1, X_2)-plane will be located inside a rectangle,
the sides of which are the two ranges. Members of Class B will
likewise be illustrated by points inside a rectangle, the sides
of which are the ranges of $_Bf_1$ and $_Bf_2$. Only in the rectan-
gular area called (AB), where the two rectangles overlap, is
there a chance of misclassifying a member of Class A as being
a member of Class B, or a member of Class B as being a mem
ber of Class A. In the manner just described we can define
the areas (AC) and (BC). The small rectangle which is part of
(AB), (AC) and (BC) is called (ABC). We decide the class mem
bership of an unlabelled pattern by computing $\left(_Ap\right)\left(_Af\left(s_1, s_2\right)\right)$,
$\left(_Bp\right)\left(_Bf\left(s_1, s_2\right)\right)$, and $\left(_cp\right)\left(_cf\left(s_1, s_2\right)\right)$; whichever class has the
highest valued product of a priori probability and density at
(s_1, s_2) is assumed to be the class from which the unlabelled
sample came; on the average this decision rule results in the
smallest number of misclassification problem may be found
in the literature $\begin{bmatrix}10, & 11\end{bmatrix}$. In the area ((AB) - (ABC)), where

$_{C}f = 0$, the average classification error is the function

$$\text{Min} \left\{ (_{A}P)(_{A}f), (_{B}P)(_{B}f) \right\}$$

integrated over $((AB) - (ABC))$.

In the area (ABC) where none of the three densities can be ex
pected to be zero, the average classification error is the func
tion Φ ,

$$\Phi = (_{A}P)(_{A}f) + (_{B}P)(_{B}f) + (_{C}P)(_{C}f) -$$

$$- \text{Max} \left\{ (_{A}P)(_{A}f), (_{B}P)(_{B}f), (_{C}P)(_{C}f) \right\},$$

integrated over (ABC). Consequently the set of three densities,
which maximizes the average classification error, in the set
which maximizes the function F.

$$F = \int_{(ABC)} \Phi d(area) + \int_{((AB)-(ABC))} \Phi d(area) + \int_{((AC)-(ABC))} \Phi d(area) + \int_{((BC)-(ABC))} \Phi d(area).$$

The largest average pattern-classification error-probability,
F_{max} , may now be found by a hill-climbing procedure, where
$_{A}f, _{B}f$ and $_{C}f$ are modified one at a time by a Θ-transforma-
tion. The above argument can be extended from the bivariate
case to the multivariate case, $N > 2$, without conceptual diffi
culties.

7. Conclusions

The main accomplishment in this paper is the introduction of the Θ-transformation. With this new transformation multivariate probability density functions can be changed while retaining their marginals. In many practical problems only the marginals are known, while some function of the multivariate density is of interest. Thanks to the Θ-transformation it has now become possible in an organized manner to search for interesting multivariate joint densities which have a specified set of marginals.

8. Acknowledgement

The author wishes to acknowledge the many helpful discussions he has had with Professor Georg Bruun, Dr. Mark Ballan and members of the staff at the Electronics Laboratory, Technical University of Denmark.

APPENDIX 1

The Generation of a String of Θ-Transformations

In this appendix we present a simple case study illustrating a difficulty encountered during the generation of a sequence of Θ-transformations. Figure 2 illustrates a simple case where $N = 3$. The parameters X_1, X_2, and X_3 all have two values: 0 or 1; n is consequently $2^3 = 8$. Let us assume that we want to transform f_p to f by a series of Θ-transformations and that the following changes are desired:

$$_{(0,0,0)}d = q \; , \qquad _{(1,1,1)}d = 2q,$$

$$_{(1,1,0)}d = _{(1,0,1)}d = _{(0,1,1)}d = -q \; .$$

In other words, we intend to take q units of probability mass from X_{f_p} at the three points (1, 1, 0), (1, 0, 1) and (0, 1, 1); then, using Θ-transformations, we intend to increase X_{f_p} by 2q at (1, 1, 1) and by q at (0, 0, 0).

It now turns out to be of some consequence which road we follow from f_p to f, meaning which sequence of Θ-transformations we select. An example with 2 sequences will clarify this point.

First sequence. First we decide to move q units of probability mass to (1, 1, 1) where 2q are needed. We move q units from

$(1, 0, 1)$ to $(1, 1, 1)$ and simultaneously q units from $(1, 1, 0)$ to $(1, 0, 0)$. By now $_0D = 6$ has decreased to $_1D = 4$, and

$$^{(1,0,0)}_1 d = {}^{(0,1,1)}_1 d = -q \, ,$$

$$^{(0,0,0)}_1 d = {}^{(1,1,1)}_1 d = q \, .$$

Next, we move q units from $(1, 0, 0)$ to $(0, 0, 0)$ and simultaneously q units from $(0, 1, 1)$ to $(1, 1, 1)$; $_1f$ has now been transformed to $_2f = f$ and $_2D = 0$. We have achieved what we set out to do.

Second Sequence. If we decide as our first move to move q units of probability mass to $(0,0,0)$, rather than to $(1,1,1)$, we find that we cannot take the mass directly from any two of the three points where there is excess probability mass. The reason why we cannot move mass directly to $(0, 0, 0)$ is that no two of the three points together have a first coordinate which is zero, a second coordinate which is zero, and a third coordinate which is zero. We realize that we cannot always move excess probability mass directly to the lattice point where it is needed. $(0, 0, 0)$ in our case. Instead, we adopt the strategy in a series of steps to move excess mass to lattice points which agree with $(0, 0, 0)$ in more and more coordinate values; this strategy will be explained more fully in Appendix 2. At this point let us introduce a useful notation. The four lattice points with $x_1 = 0$ are referred to as the set of points S_1 ; the four points with $x_2 = 0$ are referred to as set S_2 , the 2 points with $(x_1, x_2) = (0,0)$ are refer red to as the set S_{12} . The new strategy leads to the following steps. First we look for lattice points in S_1 and S_2 with excess probability mass; we find $(0, 1, 1)$ and $(1, 0, 1)$ and use them in a Θ-transformation to move q units of excess mass to $(0, 0, 1)$ which belongs to S_{12}, at the same time we also move q units to $(1, 1, 1)$. The fact that excess mass was needed at $(1, 1, 1)$ is pure ly coincidental. $_1D$ has the value four. We have succeeded in locating q units of excess mass in S_{12} where the points agree with $(0, 0, 0)$ in the values of (X_1, X_2), and we look for a point in S_3 with excess mass; points in S_3 have the same X_3-value as $(0, 0, 0)$. We find $(1, 1, 0)$ with q units of excess mass. As second step we do what we set out to do: we make a Θ-transformation by which q units are carried to $(0, 0, 0)$; at the same time it so happens that another q units are moved to $(1, 1, 1)$ and f_2 becomes f; $_2D$ becomes zero and the transformation is over.

The proposed strategy will now be described in more detail in Appendix 2.

APPENDIX 2

Generation of an " Admissible Sub-Sequence of Joint Densities"

In this appendix we will describe a particular method for generation of an admissible sub-sequence of finite length; by "finite length" we mean that the number of θ-transformations is finite. The method has the property that during the subsequence the value of the "distance measure", D, will decrease with at least 2q units of probability mass. As evidenced by Equations 3 and 4, the initial value of D, $_0$D, can never exceed two units of probability mass; the value of D can consequently be reduced to zero by generating at most $2/(2q) =$ $= 1/q$ of the finite-length sub-sequences, one after the other. In other words, f_p can be transformed to f by a finite number of θ-transformations as promised in the theorem, Section 4. Before we proceed it is necessary to introduce some additional notation. For the sake of simplicity we will in this appendix only describe how the first of the W subsequences is generated. This results in no loss of generality; the remaining (W-1) subsequences are generated in the same manner. The described method may not be particularly elegant or fast from a mathematical point of view, however, it is hoped that the method is conceptually simple. A value of $_0^x$d is associated with each of the n lattice points. We assume that $f \neq f_p$ so that at least one of the points has a $_0^x$d which is q or more positive; let such a point be $P = (a_1, a_2, \ldots, a_i, \ldots, a_N)$. The $_0^X d$-value at P is called $_0^P d$. n/n_i of the lattice points have the same value of the ith coordinates as $P, x_i = a_i, i = 1, \ldots, N$; the collection of points is referred to as the set of points S_i. Whenever the ith coordinate has values which are different from a_i, we will use the notation $x_i = \bar{a}_i$; $n/(n_1, n_2)$ points have $(x_1, x_2) = (a_1, a_2)$. We refer to the collection of these points as the set of points S_{12}; S_{12} is a subset of both S_1 and S_2. $n/(n_1 n_2 \ldots n_k)$ points have $(x_1, x_2, \ldots, x_k) = (a_1, a_2, \ldots, a_k)$. We refer to the collection of these points as the set of points $S_{12\ldots k}$; the set is a subset of $S_1, S_2, \ldots, S_{(k-1)}$ and S_k. The set $S_{12\ldots k}$ does only contain the point P; the set is, for the sake of brevity, called the set P. P is a subset of all sets mentioned in this paragraph.

We now have most of the notation needed to describe the generation of the sub-sequence. The reader can visualize the n points and their associated $_0^X d$-values. At each point X where $_0^X d$ is negative there are $-(_0^X d)$ units of excess probability mass which we want to remove from the point; at each point where $_0^X d$ is positive there is needed $(_0^X d)$ units of excess probability mass. The algebraic sum of $_k^X d$ for the members of any set S_i is zero because the value of the marginal remains $f_i(a_i)$ during θ-transformations. The sole objective of the sub-sequence is to move $(_0^P d)$ units of excess probability mass to P and thereby reduce the value of $^P d$ to zero or some negative number; the sub-sequence consists of $(N-1)$ steps.

The first step. We are concerned with the set S_{12} and the two difference sets $(S_1 - S_{12})$ and $(S_2 - S_{12})$. We check the two difference sets for points with negative-valued $^x_0 d$, meaning points with excess probability mass; let the sum of all excess mass associated with the points in $(S_1 - S_{12})$ be E_1, and let likewise the total excess mass associated with points in $(S_2 - S_{12})$ be E_2, where E_1 and/or E_2 could very well be zero. Whenever we have found a point with excess mass in $(S_1 - S_{12})$ and a point with excess mass in $(S_2 - S_{12})$ we use the two points as P_1 and P_2 in a θ-transformation. We transfer excess mass to some point in S_{12} which functions as P_3; at the same time we transfer an equal amount of excess mass to some point which has $(x_1, x_2) = (\bar{a}_1, \bar{a}_2)$; this point functions as P_4. After we have moved $\text{Min}\{E_1, E_2\}$ excess mass by a series of θ-transformations, one (or both) of the two difference sets have no more points left with excess mass; this means that: at the end of the first step the total amount of excess mass in S_{12} is at least as large as the present value of $^P d$. If the first step involved z θ-transformations, the present value of $^P d$ is $^P_z d$; $^P_z d$ will usually be positive. Also $^P_z d$ could be zero or negative, if so we did already reach our objective when the value of $^P d$ became non-positive and we could have stopped right there; if we nevertheless continue, the following steps are unnecessary though no damage is done by taking them as the value of D will not increase. The number of points in $(S_1 - S_{12})$ and $(S_2 - S_{12})$ are

$(n / n_1 - n / (n_1 n_2))$ and $(n / n_2 - n / (n_1 n_2))$; z is bounded by their
product and consequently by $(n / n_1)(n / n_2)$, $z < n^2 / (n_1 n_2)$.
In the next (N-2) steps we will essentially repeat what was done
in the first step.

The second step. We are concerned with the set S_{123} and the
two difference sets $(S_{12} - S_{123})$ and $(S_3 - S_{123})$; the three sets will
play the same part in step n° 2 as S_{12}, $(S_1 - S_{12})$ and $(S_2 - S_{12})$
played in step n° 1. The sum of all excess probability mass in
$(S_{12} - S_{123})$ after θ-transformation n° z is called E_{12} ; the total
excess mass in $(S_3 - S_{123})$ is called E_3. Before the second step
is initiated the sum of the excess mass associated with points
in S_{123} may well be less than $\overset{P}{z}d$. As before we perform θ-
transformations where we use points in $(S_{12} - S_{123})$ as P_1 and
points in $(S_3 - S_{123})$ as P_2. Points in $(S_{12} - S_{123})$ have $(x_1, x_2, x_3) =$
$= (a_1, a_2, \bar{a}_3)$. Points in $(S_3 - S_{123})$ have $(x_1, x_2, x_3) = (\bar{a}_1, \bar{a}_2, a_3)$ or
$(\bar{a}_1, a_2, a_3)$ or (a_1, a_2, a_3). Points in S_{123} we use as P_3, they
have $(x_1, x_2, x_3) = (a_1, a_2, a_3)$. Points in the set where $(x_1, x_2, x_3) =$
$= (\bar{a}_1, \bar{a}_2, \bar{a}_3)$ or $(a_1, \bar{a}_2, \bar{a}_3)$ or $(\bar{a}_1, a_2, \bar{a}_3)$ we use as P_4. Whenever
ever a total of $\mathrm{Min}\{E_{12}, E_3\}$ mass has been moved to points in
S_{123}, one of the two difference sets is depleted of points with
excess mass (the case where they are simultaneously depleted,
$E_{12} = E_3$, only gives rise to a trivial extension to the following discussion). Let us consider the two possibilities one at a
time.

(1) S_3 is the union of the disjoint sets $[12]$ S_{123} and $(S_3 - S_{123})$.

If $(S_3 - S_{123})$ is first depleted of excess mass the algebraic sum of ^Xd-values for members of S_{123} is negative or zero; the value of ^Pd at the end of step n° 2 is consequently equal to or less than the sum of excess mass in S_{123}. (2) as pointed out before, the algebraic sum of ^Xd values for points in S_{12} is zero or negative; during the second step the value of the sum remains unchanged. S_{12} is the union of the disjoint sets $(S_{12} - S_{123})$ and S_{123}. If therefore $(S_{12} - S_{123})$ is first depleted of excess mass the algebraic sum of ^Xd-values for members of S_{123} is negative or zero; also in this case it is seen that: at the end of the second step the value of ^Pd is equal to or less than the sum of excess mass associated with points in S_{123}.

The kth step. Having discussed steps n° 1 and 2 in detail, we now turn to the general case, step n° k. Let us assume that we at the end of step n° (k-1) were able to demonstrate that: "at the end of step n° (k-1) the value of ^Pd is equal to or less than the sum of excess mass in $S_{12...k}$"; we have already demonstrated the statement for k = 2 and k = 3 in the first two steps. If so we can prove the statement in the next sentence by following the procedure from "the second step" and substituting the sets $S_{12...k}$, $S_{(k+1)}$, and $S_{12...(1+k)}$, for the sets S_{12}, S_3, and S_{123}; we also substitute $E_{12...k}$ and $E_{(k+1)}$ for E_{12} and E_3. At the end of step n° k the value of ^Pd is equal to or less than the sum of excess mass associated with points in

$S_{12...(k+1)}$.

We now follow the line of reasoning familiar from theorem proving by induction. In "the second step" we showed that the above assumption is true for $k = 2$, but thanks to the argument in this paragraph the underlined statement above becomes true for $k = 3$. However, by repeating the argument in this paragraph for the case $k = 4$ we can show that the underlined statement actually is true for $k = 4$. We can repeat the procedure for all values of k up to and including $k = N-1$. Step n° $(N-1)$, where $k = N-1$, is the last step and deserves some comments. The last step meaning step n° $(N-1)$. According to the last underlined statement we have that "at the end of step n°$(N-1)$ the value of $^P d$ is equal to or less than the sum of excess mass associated with the points in $S_{12...N}$ ". $S_{12...N'}$, which is also called the set P, does only contain the point P. All excess mass which was moved to the set during step n° $(N-1)$ was therefore located at the point P where it satisfied any remaining need for probability mass; only after step n° $(N-1)$ can it be stated with certainty that the value of $^P d$ is non-positive.

Results. Let us check that what was promised at the end of Section 5 actually has been achieved. We have only moved excess mass so that after each θ-transformation the value of D stayed constant or decreased; this matter was discussed in the material leading up to Equation 5. (1) The number of θ-transformations used in the first step is less than $(n/n_1)(n/n_2)$, the number used in the kth step is less than $(n_{k+1} \ n_{k+2} \cdots n_N)(n/n_{k+1})$, the number used in the last step is less than n ; <u>the number of Θ-transformations used in the sub-sequence is consequently finite</u> and bounded by the N values, $n_1, n_2, \ldots, n_N$. (2) The fact that $\binom{P}{0}d$ units of excess mass were transferred to P insures that the value of D has decreased by at least $2\binom{P}{0}d$. Half of the decrease has to do with the change of the ${}^P d$ -value from ${}_0^P d$ to zero. The other half of the decrease is due to the fact that the need for mass at P was satisfied with excess mass from other points; at these other points the (negative) d -values increased with a total of ${}_0^P d$ in the process. During the generation of the sub-sequence some excess mass may by coincidence have been transferred to points where there was a need for excess mass $\left[$ the d -value was positive $\right]$; if so the value of D will have d<u>e</u>creased by more than $2\binom{P}{0}d$. ${}_0^P d$ is q or more; the value of D will consequently decrease with at least $2q$ during a sub-s<u>e</u>quence of the described type.

REFERENCES

[1] Jones, Bush and Brigham, Oran: "A Note on the Deter
mination of a Discrete Probability Distribution
from Known Marginals", Information and Con
trol vol. 15 n°6, pp. 527-528, December 1969.

[2] Strassen V.: "The Existence of Probability Measures
with Given Marginals" The Annals of Mathe
matical Statistics, vol. 36, pp. 423-439, 1965.

[3] Kullbach S.: "Probability Densities with given Margin
als", The Annals of Mathematical Statistics,
vol. 39, pp. 1236-1243, 1968.

[4] Papoulis Athanasios: "Probabilities, Random Variables
and Stochastic Processes", New York, McGraw
-Hill, 1965; Figures 6-19 and 7-20.

[5] Fine Terence L.: "On the Apparent Convergence of Rel
ative Frequency and Its Implications", IEEE
Trans. on Information Theory, vol. IT-16,
n° 3, pp. 251-257, May 1970.

[6] Papoulis Athanasios, op. cit.; p. 8.

[7] Ibid., Section 8-1.

[8] Becker P.W.: "Static Design of Transistor Diode Logic"
IRE Trans. on Circuit Theory, vol. CT-8,
n° 4, pp. 461-467, December 1961.

[9] Becker P.W.: "Graphs Speed Unsymmetrical Flip-Flop
Design", Electronic Design, Sept. 13 and 27,
1963.

[10] Becker P.W.: "Recognition of Patterns", Copenhagen
Denmark, Polyteknisk Forlag, 1968; Sect. 1.6.

[11] Nagy G.: "State of the Art in Pattern Recognition" Proc.
            IEEE, vol. 56 n°5, pp. 836-862, May 1968 .

[12] Papoulis Athanasios, op. cit. ; Section 2-1.

CONTENTS

Page

PREFACE 5

GLOSSARY OF SYMBOLS 7

Chapter I.   Introduction 11

    1.1   The Purpose of the Paper 11
    1.2   The Two Phases in the Existence
                of a PRD 12
    1.3   Three Areas of Application 13
    1.3.1   Certain Acts of Identification ... 13
    1.3.2   Decisions in Regarding Complex
                Situations 15
    1.3.3   Imitation of Human Pattern Rec-
                ognition 16

Chapter II.   The Configuration of a PRD 16

Chapter III.   A Step by Step Description of the
                the PRD Design Procedure 20

Chapter IV.   Factors that Influence the Design
                of a PRD 29

    4.1   Factors Associated with the Pat-
                tern Classes 29
    4.2   Estimation of a PRD's Perfor-
                mance 32
    4.3   Four Major Problem Areas 34

Chapter V.   The Selection of the Attributes . 35

    5.1   Preliminary Processing 35
    5.2   Generation of Sets of Attributes
                in Practice 37

			Page
5.3	An Effective Set of Attributes ...		39
5.3.1	A Definition of "An Effective Set of Attributes"		39
5.3.2	A Definition of "The Incremental Effectiveness of an Attribute"...		42
5.4	One Attribute		42
5.5	Templet Matching		47
5.6	Selection of a Set of P Attributes		50
Chapter VI.	Decision Procedures and Indices of Performance		55
6.1	Some Functions Related to the Micro-Regions		56
6.2	Bayes' Procedure		58
6.3	The Minimaxing Classification Procedure		59
6.4	The Likelihood Method		61
6.5	The Neyman-Pearson Method ..		63
6.6	Three Practical Difficulties ...		64
Chapter VII.	Categorizer Design		66
7.1	Estimation of a Multivariate Density Function		67
7.2	Explicit Partitioning of the Pattern Space		71
7.2.1	Separation Surface of Simple Shape		71
7.2.2	The Need for Separation Surfaces of Simple Shape		73
7.2.3	Parametric Training Methods ..		77
7.2.4	Non-Parametric Training Methods		80
7.3	Implicit Partitioning of the Pattern Space		84
7.3.1	Nearest Neighbour Pattern Classification		84

		Page
7.3.2	Discriminant Functions and Separation Surfaces	85
7.3.3	Categorization Using N_c Discriminants	87
7.3.4	The Φ Machine	88
7.3.5	The Nonlinear Parametric Discriminant	89
7.3.6	Parametric Training of Discrim inants	90
7.4	Categorization of Members from More than Two Classes .	92
Chapter VIII.	Hardware Implementation	97
APPENDIX I.	....	101
1.	Introduction	101
2.	A Topic for a Masters Thesis.	101
3.	Recognition of Handprinted Digits	102
4.	The Recognition Logic	103
5.	The Dynamic Attributes	104
6.	The Static Attributes	105
7.	The Decision Functions	106
8.	The Hardware Implementation	107
APPENDIX II.	....	111
APPENDIX III.	....	113
1.	The Performance of the PRD.	113
2.	The Design of the PRD	114
APPENDIX IV.	....	117
1.	Introduction	117
2.	Some Considerations	117
BIBLIOGRAPHY	....	121

ADDENDUM

                                                    Page

ABSTRACT   147

GLOSSARY OF SYMBOLS   149

        1.   Introduction   151

        2.   Some Preliminary Assumptions   152

        3.   The Θ -Transformation   154

        4.   A Theorem   155

        5.   A Proof of the Theorem   157

        6.   The Case of Continuous Den-
                  sities   161

        7.   Conclusions   169

        8.   Acknowledgement   169

APPENDIX I.   171

APPENDIX II.   175

REFERENCES   183

CONTENTS   185

Printed in the United States
By Bookmasters